It's All in the Genes! Really?

Dr. Gerard M. Verschuuren

ISBN: 1496031687
ISBN-13: 978-1496031686

DEDICATION

Dedicated to all those who think
they were given the "wrong genes."

ACKNOWLEDGMENTS

I wish to express my gratitude to those nameless who steered and corrected me in the process of preparation for this book. In particular, I want to mention Ronald Arellano, M.D. (Massachusetts General Hospital), Francisco Ayala (University of California at Irvine), Stephen Barr (University of Delaware), Paul Camarata, M.D. (University of Kansas School of Medicine), Daniel Fairbanks (Utah Valley University), Uta Francke (Stanford University), and Peter Kreeft (Boston College).

And—as with all my English books—I could not have written this book without the unwavering and loving support of my wife, Trudy. I also want to specifically mention the great and helpful comments of our daughter Tamson Jane Kelley, who works in the medical field and helped me in polishing the text.

They and many others make me realize that originality often consists in the capacity of forgetting about your sources. Obviously, they are not responsible for the outcome—if I erred, it's entirely my doing.

Table of Contents

TABLE OF CONTENTS...II

ENDORSEMENTS..III

ABOUT THE AUTHOR V

PREFACE VI

I. BEHIND THE GENES.. 1

    1. THE HISTORY OF THE GENE 1
    2. THE POWER OF GENES 8
    3. IS IT ALL IN THE GENES? 19
    4. NATURE VERSUS NURTURE 30

II. THE BATTLE OF GENES 39

    5. THE GENE POOL 39
    6. NATURAL SELECTION 49
    7. WHAT SELECTS WHAT? 55
    8. CAN THINGS GO WRONG? 60

III. THE DOUBLE HELIX 71

    9. FROM GENES TO DNA 71
    10. CHANGES IN DNA 82
    11. THE SILENT PARTS OF DNA 93
    12. DNA, THE SECRET OF LIFE? 103

IV. WHAT NO GENE CAN DO... 111

    13. NO GENE FOR FREE WILL 111
    14. NO GENE FOR RATIONALITY 127
    15. NO GENE FOR MORALITY 144
    16. NO GENE FOR RELIGION 160
    17. WHEN DID IT START? 175

V. CONCLUSION 183

VI. INDEX.. 187

Endorsements

"Where does morality come from? What about religion and rationality? Do we really have free will? If you want answers to these questions, you will want to read It's All in the Genes! Really? *You will also learn much about genetics, genomics, DNA and much more, all in a style that is informative, clear and enjoyable."*

Francisco J. Ayala is University Professor and Donald Bren Professor of Biological Sciences at the University of California, Irvine. Department of Ecology and Evolutionary Genetics.

————————————————

"The critical discussion by Dr. G.M. Verschuuren of selected examples for acquired insights into the complexity of life represents a welcome contribution to inform a wider public on questions and at least partial answers dealt with by fundamental and applied research in the life sciences".

Professor Werner Arber, Nobel Laureate in Medicine 1978, Biozentrum, Department of Microbiology, University of Basel, Switzerland.

————————————————

"There is no such thing as a perfect genome. We always knew. But in the era of exome and genome sequencing, we can actually see it, right in front of our own eyes. Each one of us carries hundreds to thousands of variants or mutations. What a humbling experience. The consequences of most of these we don't yet understand. And then, beyond biology, there is more. What makes us human? Where does biology end, and where does philosophy begin? Gerard Verschuuren presents an intelligent book, which will educate, stimulate, provoke, and inspire."

About the Author

Dr. Gerard M. Verschuuren is a human geneticist who also earned a doctorate in the philosophy of science. He studied and worked at universities in Europe and the United States and wrote several biology textbooks in Dutch. Currently, he is semi-retired and spends most of his time as a writer, speaker, and consultant on the interface of science and religion, creation and evolution, faith and reason.

His most recent books are:

- Darwin's Philosophical Legacy—The Good and the Not-So-Good (Lanham, MD: Lexington Books, 2012).
- God and Evolution?—Science Meets Faith (Boston, MA: Pauline Books, 2012).
- Of All That Is, Seen and Unseen—Life-Saving Answers to Life-Size Questions (Goleta, CA: Queenship Publishing, 2012).
- What Makes You Tick?—A New Paradigm for Neuroscience (Antioch, CA: Solas Press, 2012).
- The Destiny of the Universe—In Pursuit of the Great Unknown (St. Paul, MN: Paragon House, 2014).
- Life's Journey—A Guide from Womb to Tomb (upcoming, Goleta, CA: Queenship Publishing).

For more info see:

http://en.wikipedia.org/wiki/Gerard_Verschuuren

He can be contacted at www.where-do-we-come-from.com

Preface

"It's all in the genes" is one of those statements that you may hear almost daily on radio and TV, on the internet, in your own circle of friends and colleagues. It is one of those explain-it-all statements. I used it myself for a while when studying and later teaching human genetics. Like all other geneticists, I was brought up with the prevailing paradigm of genetics. The term paradigm, which Thomas Kuhn introduced, stands for a collection of rules on how to solve scientific puzzles—in this case, genetic puzzles.

Most scientists feel attached to the paradigm they were brought up with. The reason is that individual scientists acquire knowledge of a paradigm through their scientific education. That is how they learn their standards, by solving "standard" problems, performing "standard" experiments, and eventually by doing research under a supervisor already skilled within the paradigm. Aspiring scientists become gradually acquainted with the methods, the techniques, and the presuppositions of that particular paradigm. And geneticists are no exception.

Because of their training, scientists are typically unable to articulate the precise nature of the paradigm in which they work, until a need arises to become aware of the general laws, simplified models, metaphysical assumptions, and methodological principles involved in their paradigm. I think I have reached a turning point somewhere in my life. That is when I changed my motto of "It's *all* in the genes" into "It's in the genes, but there is more than genes in life."

In this book, I will give fair attention to both parts of my newly discovered motto. On the one hand, I will show how powerful genes, and their DNA, can be—for we need to acknowledge that "It's in the genes." On the other hand, I will also explain why genes do not tell the entire story—for we should also recognize that "there is more than genes in life, so genes do not answer *all* our questions." That double focus explains why the outline of this book is the way it is.

In part I, "Behind the Genes," we discuss what we currently know about genes. It is more or less an overview of "classical" genetics. This is a rather basic section. I start with how the gene concept came along (Chapter 1). Then I explain how genes can have so much power (Chapter 2). This calls inevitably for a discussion of where their power ends (Chapter 3). Finally, I delve into the famous nature-nurture debate (Chapter 4).

In part II, "The Battle of Genes," I discuss how genes can cause differences between people. First, we talk about the gene-pool model and how its composition can change over time (Chapter 5). This takes us into the discussion as to what natural selection did and still does to human populations (Chapter 6). This might stir some controversy as to what actually selects what during natural selection (Chapter 7). Then I would like to discuss whether and how "things can go wrong" (Chapter 8).

In part III, "The Double Helix," we go into the mechanism behind the genes—the workings of DNA. This is probably the most technical section of the book, but I tried to explain every detail as easily and clearly as possible, without going overboard. First, I discuss how DNA does its job (Chapter 9). Then we need to find out how DNA could change over time, which has something to do with the various kinds of mutations genetic material may undergo (Chapter 10). Since there are large parts of DNA that do not contain genes, we need to discuss also what those parts might be doing, and how they may help us determine where we came from (Chapter 11). Finally, we need to put DNA in its proper place—what it can and cannot do, and how it came into existence (Chapter 12).

In part IV, "What No Gene Can Do," I try to put genetics back into its proper place by showing what it can*not* do. This is probably the most controversial section of the book, for I use science, logic, and philosophy to show that it is pointless to go in pursuit of genes for our free will (Chapter 13, touching also on addictions), of genes for our human capacity of rationality (Chapter 14, in comparison with animal intelligence), of genes for our human capacity of morality (Chapter 15, including the sociobiology of altruism), or of genes for our belief in God (Chapter 16, including the god-gene hypothesis). I do all of this in spite of often heard claims to the contrary. Finally, I go into the discussion if genetics can help us determine when and where humanity first appeared on the world's scene (Chapter 17).

When I was teaching philosophy of biology at Boston College, the chairman of my department at the time, Joseph Flanagan, S.J. instilled in me that biology can only fare well with the right philosophy. I am sure he would have said something similar about genetics. That is why philosophy makes up an integral part of this book.

I decided not to use footnotes with the sources of my data—so as to keep the book readable for the more general reader. Yet be assured that, as a human geneticist, I tried to only select information that is reliable and has been corroborated. I am not saying this is also definitive information, as there is never final certainty in science. As Francis Crick, one of the two scientists who discovered DNA, put it, "A theory that fits all the facts is bound to be wrong, as some of the facts will be wrong." Scientific information is very volatile by its very nature— always a work-in-progress. When this book comes off the press, certain information may already be outdated. Reader, be aware.

I. Behind the Genes

1. The History of the Gene

When you ask geneticists nowadays who was the first one to talk about genes, they almost certainly come up with the name of Gregor Mendel, who lived from 1822 until 1884. Mendel may have been practically unknown during his lifetime, but thanks to some geneticists after him, he has been heralded as "the father of genetics."

For almost 35 years, Mendel's findings escaped notice, until other geneticists such as Hugo De Vries in Holland, Erich Von Tschermak in Austria, Carl Correns in Germany, and William Bateson in England found similar data. When they came across Mendel's original paper from 1866, this helped them interpret their own data better. Bateson probably learned of Mendel from De Vries. Correns and Bateson, in particular, were fair enough to pay homage to Mendel, and because of this, he has been coined the founder of what became known as "Mendelism."

How was it possible that Mendel was not noticed for such a long time? There are many possible explanations. Perhaps Mendel was operating on the fringes of the scientific community and published his article in an unknown, rather provincial journal. Then there is the possibility that rivalry and jealousy are to blame, for Mendel did send copies of his article to many colleagues, but they hardly reacted. Or perhaps Mendel's discoveries were "premature," because the time was not "ripe" yet for such revolutionary ideas. All these explanations are possible, but I am afraid these are not the main reasons. There was something else going on.

It seems much more likely that Mendel was not really interested in the question of how the transmission of hereditary factors is achieved. Nowadays, we would say he was part of a different research tradition, a different paradigm, or a different research program. The paradigm of genetics centers on the transmission of hereditary factors— currently called genes. But that issue was not Mendel's main concern. Mendel was still part of an older research tradition dating back to Carl Linnaeus (1707-1778) which centered on the question as to whether hybrids are really able to form a new species. Later in life, Linnaeus had been led to believe that perhaps only genera had been created in the beginning and that species were the product of hybridization within these genera.

In the eighteenth century, Joseph Kölreuter had already shown in a series of experiments that newly produced hybrids between species are not *constant* new species but could be returned to the parental species by continuous back-crossing. Interestingly enough, these experiments were explicitly mentioned in Mendel's articles. However, the question of whether hybrids are really able to form a new species is not an issue for geneticists but for cross-breeders—a mixture of species-breeders and plant-breeders. They come from a different research tradition and work with a different paradigm.

At first sight, it looks as if Mendel is really talking genetics. He did his experiments with garden peas and used seven different pairs of contrasting traits—such as seed shape (round vs. wrinkled) and stem length (short vs. tall). This could be something he had learned from plant-breeders. When plants with round peas were crossed with plants with wrinkled peas, all the offspring produced round seeds. Because one trait seemed to "dominate" over the other trait, Mendel called such traits *dominant*. When he let these new plants pollinate themselves, he found that 75% produced round seeds, but 25% wrinkled seeds again—which is a ratio of 3:1. In other words, the characteristic of wrinkled seeds had receded only temporarily, so he called that trait *recessive*. Does all of this not remind you of your first lessons in genetics?

Nevertheless, it is very doubtful whether Mendel was talking genetics here. He comes deceivingly close to saying *AA x aa* produces 100% *Aa*, whereas *Aa x Aa* produces 25% *AA*, 25% *aa*, and 50% *Aa*. You probably think that *AA* and *aa* are *homozygotes*, whereas *Aa* organisms should be called *heterozygotes*. But that impression is actually incorrect. In fact, Mendel symbolizes constant forms—which we would now call homozygotes—by just *one* letter (*A* or *a* instead of *AA* or *aa*). Hybrids, on the other hand—which we would now call heterozygotes—he did express in our "modern" notation using *two* letters (*Aa*).

It is very important to notice that Mendel is very consistent in characterizing constant "forms" with *one* letter, and only the hybrids (as not being constant) with *two* letters. In Mendel's view, an *Aa*-organism is a hybrid, not a heterozygote. A hybrid is a breed of two pure types and carries two different "elements" from each type. Even germ cells he called "of form A."

I think it is hard to deny that Mendel was in search of the laws of hybridization, not the laws of genetics. Mendel's concept of hybrid is still part of the hybridization theories of the 19th century, and not of the hereditarian theories of the 20th century. Inspired by what species-breeders did, by doing his experiments, Mendel found out that, after one generation, a cross between two pure forms would yield these original pure forms again. His discovery proved that there is an essential difference in those hybrids which remain constant in their progeny and propagate themselves as truly as the pure species.

It is hard to deny that Mendel was more interested in the constancy of species than in the heredity of characteristics. He was inspired by what species-breeders were searching for. So by reading Mendel through the eyes of later geneticists, he comes out a bit distorted. Translating his ideas into concepts we know from genetics—as Bateson did—is a bit unfitting. So speaking of "Mendelian laws" is not quite accurate, yet very common.

Nevertheless, Mendel did things future geneticists would admire him for—and rightly so. First of all, instead of studying only the few offspring from a single mating, he studied the offspring of many pairs of similar parents, so he could treat them as if they had resulted from one single mating. Second, since he was dealing with large numbers of offspring, Mendel could use mathematics, more in particular probability calculus—the mathematical science that tries to predict the chances that a certain event may occur. In short, he was able to make predictions and give explanations by using statistical laws. Third, he did not try to study everything about all the offspring at once. Instead, he selected only a few individual traits of pea plants and studied them in detail, which he had probably learned from plant breeders. Yet, the combination of all of this made Mendel unique.

No wonder, Correns and Bateson got the impression Mendel had done already in 1865 what they were doing in 1900. The adjective "genetic" was already used by Huxley as early as 1864, but Bateson adapted this nineteenth century term to inaugurate a new science, which he named "genetics." He also phrased the new terminology of genetics and applied it to Mendel's findings. According to this new terminology, Mendel had been working with two forms of the *gene* for seed shape—one that produced round peas and another one that produced wrinkled peas. From now on, when a gene exists in more than one form, the different forms are called *alleles*. When an organism contains two of the same alleles, it is called *homozygous*; if it contains two different alleles of the same gene, it is called *heterozygous*. If one allele is dominant (*A*) over the recessive one (*a*), visual inspection does not tell us whether the organism is homozygous for the dominant allele (*AA*) or heterozygous (*Aa*). So one particular *phenotype*—say, round peas—can have different *genotypes*—say, being homozygous for round or heterozygous for round-and-wrinkled.

The same concepts can be applied to situations where we have two or more pairs of genes. Mendel had already discovered that a cross between plants with round-yellow peas and plants with wrinkled-green peas results in 100% round-yellow seeds, but when he allowed them to naturally self-fertilize, he found 9:3:3:1 for round-yellow, wrinkled-yellow, round-green, and wrinkled-green respectively. Apparently, the genes for these two characters get passed on to the new generation *independently* of each other. So far, that sounds like solid "classical" genetics.

We should realize, though, that at the beginning of the 1900s it was still unclear what the nature of these genes and alleles was. They were still entities of a rather hypothetical nature. What Mendel had referred to as "elements" and Bateson as "factors" was named "genes" by Wilhelm Johanssen, but they were still taken as some sort of accounting or calculating unit.

Things began to change when geneticists discovered that two or more pairs of genes may not *always* separate independently. Sometimes genes seem to be more or less *linked* to each other. It had already been noticed by the geneticist Walter Sutton that genes always occur in pairs and that chromosomes also occur in pairs. Was this mere coincidence? Or could it be that genes and chromosomes are connected, and that genes are actually located in chromosomes?

Humans, for instance, have only 23 pairs of chromosomes, but they have thousands of different genes. If there is a connection between genes and chromosomes, then there must be many different genes per chromosome. If so, genes located on the same chromosome cannot separate independently but must be "linked," whereas genes located on different chromosomes can still separate independently of each other.

At last, a physical connection had been made! The next discovery was that linked genes—that is, those residing close to each other in the same chromosome—can still separate independently to a certain degree, but never for 100%. This could be explained by the fact that segments of paired chromosomes can undergo "crossing-over" during the formation of reproductive cells, thereby exchanging segments of the chromosome. Although the pea plant *Pisum sativum* of a plant breeder had been replaced by the fruit fly *Drosophila melanogaster* of a geneticist, the results were basically the same.

Yet, acceptance of chromosomes as carriers of genes remained controversial. The geneticist Thomas Hunt Morgan rejected the chromosome theory of heredity, but then embraced it wholeheartedly in his classic 1910 paper in *Science*. Finally, Mendel's hypothetical elements had found a material basis. The chromosome theory could at least explain why the gene model works.

If we assume that the probability of breakage is approximately equal at any point along the length of the chromosome, then the frequency of crossing-over between any two linked genes should be proportional to the distance between them, because there are more points between them at which the break may occur—more in general, the farther apart the more crossing-over. The percentage of crossing-over, also called *recombination*, can therefore serve as a tool for mapping the locations of genes on chromosomes. In general, it turned out that the genetic map length for all chromosomes does correspond roughly with their relative chromosome length.

A chromosome map represents the genes as lying in a line—like beads on a string, as Morgan put it. From now on, the concept of recombination became pivotal. If two characters are always linked, and crossing-over never occurs, then we assume that they are being controlled by the same gene. If, on the other hand, crossing-over does occur between them, and they can be recombined, then we assume that they must be controlled by different genes. From this follows the following definition of a gene (commonly attributed to Morgan): A gene is the smallest unit of recombination.

Later on (Chapter 9) we will see how the *Human Genome Project* envisioned to map DNA is an endeavor similar to Morgan's enterprise to create gene maps for each chromosome.

2. The Power of Genes

Although there are several more sophisticated definitions of a gene, which we will discuss later, for now I will stick with a classical definition stating that a gene is a unit of heredity that regulates a specific trait, feature, or characteristic of an organism. This is probably the most practical definition when geneticists speak of genes for tall plants or wrinkled seeds in garden peas, or of genes for forked bristles or vestigial wings in fruit flies, or of genes for curly hair or attached ear lobes in humans. But in order to decide whether two different characteristics are determined by one or two genes, we still need the more technical definition of Morgan-type genes: A gene is the smallest unit of recombination.

This is also the case in human genetics—on which we will focus from now on. Human geneticists try to explain differences between people by differences in alleles of a specific gene. They are in search of a difference that makes a difference—a difference in phenotype associated with a difference in genotype. We all know more or less intuitively that there are indeed many genetic differences between human beings. For centuries, people used to think that differences in temperament, talents, social status, wealth, and power had to reside "in the blood." They used to speak in terms of "blood" ties; terms such as "blood relative," "bloodline," "full-blooded," and "royal blood" are relics of this idea. But with the rise of Mendelian genetics, genes were substituted for blood in the explanations.

Let me make clear, though, that all humans have in essence the same genes, but not the same alleles—and differences in alleles supposedly make them different from each other. Humanity is united—yet, without losing any of the richness of variety. Human beings carry 23 pairs of chromosomes—of which one pair is an "unmatched" pair of sex chromosomes in males (*XY*), but a "real" pair (*XX*) in females. For procreation it is necessary that males produce sperm cells with a half, un-paired set of 23 chromosomes and that females release egg cells, again with a half, unpaired set of 23 chromosomes. When a sperm cell, containing only one set of 23 chromosomes, and an egg cell, con-taining also one set of 23 chromosomes, come together, they fuse their chromosomes and start a new organism with two sets of chro-mosomes again, which is 23 pairs. The Mendelian laws of inheritance explain which alleles were passed on via those chromosomes (includ-ing the possibility of recombination). From there on, the genes can do their work.

Very early in the history of genetics, the so-called "Mendelian laws" were applied to human blood types and blood groups. Currently we know of thirty different blood group systems, such as the ABO system (with a gene located on chromosome 9), the Rhesus system (on chro-mosome 1), the MSN system (on chromosome 4), the Kell system (on chromosome 7), the Lewis system (on chromosome 19), and several more. They all determine which antigens are expressed on the red blood cell surface membrane. Since most blood groups show very little variability in human populations, blood transfusion complications have been very rare, except for the ABO and Rhesus systems.

In 1900, the Austrian scientist Karl Landsteiner discovered the ABO blood group system, based on four blood types—A, B, AB, and O. Very soon after, it was shown that the ABO blood group system is based on one gene with at least three different alleles—I^A, I^B, and i. Both I^A and I^B are dominant over i, but neither I^A nor I^B is dominant over the other. Consequently, a person with the phenotype of blood type A can have two different genotypes: I^A/I^A or I^A/i. (I use here a classical notation because it is still very common in medical literature and we have not discussed yet the DNA structure of genes (see Chapter 9). The new notation for I^A would be: transferase A or alpha 1-3-N-acetylgalactosaminyltransferase (A3GALNT); and for I^B: transferase B or alpha 1-3-galactosyltransferase (A3GALT1). The i allele differs from

I^A by deletion of guanine at position 261. The deletion causes a frame shift and results in translation of an almost entirely different protein that lacks enzymatic activity.)

It speaks for itself that blood typing can be used and has been used for paternity tests in court. A man with blood type AB could not possibly be the father of a type O child, because that child must have received an i allele from its father, but an AB man has no such allele. Why is this blood type system so important, even outside the court system? Well, people with blood type O have neither the A nor the B antigen, so they attack such antigens with antibodies in their blood plasma when they are exposed to them through incompatible blood transfusions. However, this classical test is no longer widely used because DNA testing is more reliable, and is based on more than just one marker. The Bombay phenotype, for instance, would be tested as type O when in fact the child has inherited either the A or B allele.

The Rhesus system has similar consequences, but is more complicated. Individuals are classified as Rh-positive and Rh-negative according to the presence or the absence of the major D antigen on the surface of their erythrocytes, but more than 46 other antigens, including those of the *CcEe* series, have been identified. The Rh locus is composed of 2 homologous structural genes, one encoding the Rh D polypeptide and the other encoding both the Cc and the Ee polypeptides. The short, simplified story is that individuals either have or do not have the D antigen on the surface of their red blood cells—also called the "Rhesus factor." Their status is usually indicated as Rh-positive (Rh^+ has the D antigen through the dominant allele D) or Rh negative (Rh^- does not have the D antigen). However, the Rhesus blood group system is much more complicated than this, and still rather unclear, but that is beyond the scope of this book.

In daily life, Rhesus incompatibility can have serious consequences during pregnancy. When a Rh-negative woman (let us simplify her genotype as *dd*) carries the child of a Rh-positive father, then the child has either a 100% chance of being Rh-positive, if the father had *DD*, or a 50% chance, if the father had *Dd*. Usually this causes no trouble, since there is no direct connection between the two circulatory systems of mother and unborn child, and thus no mixing of blood. But in the late stages of some pregnancies and during the birth process, some seepage of blood between the two circulatory systems may occur. If the blood of a Rh-positive unborn baby seeps into the blood of a Rh-negative mother, her body is stimulated to begin synthetizing Rh antibodies. The baby is probably born before enough harmful antibodies have been produced, but if the sensitized mother later bears another Rh-positive baby, her existing Rh antibodies may seep into the baby's circulation and react with the Rh antigen of its red blood cells, which can be fatal to the baby if not treated properly. Nowadays, however, Rh sensitization in nearly all cases of Rh conflict are routinely prevented through Rhogam administration, which is an injection of Rh0 immunoglobulin to prevent the formation of antibodies to Rh positive blood.

Of course, it is not only blood group systems that follow Mendelian rules of inheritance. There are some serious diseases that can emerge when certain genes carry altered alleles. Several diseases are traceable to a single gene. Examples are Huntington disease and Tay-Sachs disease—both of which cause progressive mental deterioration. What these diseases have in common is that they follow a simplified formula: One gene can lead to one specific disease if it carries an altered allele. Let us discuss both of them in some more detail—also to get a better feel for genetics.

Huntington disease (HD) is a neurodegenerative genetic disorder that affects muscle coordination and leads to cognitive decline and psychiatric problems. Physical symptoms of Huntington's disease can begin at any age from infancy to old age, but usually begin between 35 and

45 years of age. It is much more common in people of Western European descent than in those of Asian or African ancestry. The disease is caused by a dominant allele of a gene called Huntingtin (HTT), located on chromosome 4. This gene provides the genetic information for a protein that is also called "huntingtin" (Htt). When this gene carries an altered allele, it produces an altered protein. The behavior of the altered protein is not completely understood, but it is toxic to certain types of cells, particularly in the brain. Regions of the brain have different amounts of these types of neurons, and are thus affected accordingly.

Because the altered allele is *dominant*, any child of an affected person typically has at least a 50% chance of inheriting the disease. Since the disease usually shows at a later age, some people want to be tested for the presence of the altered allele. The main reason given for choosing a test for HD is to aid in career and family decisions. Over 95% of individuals at risk of inheriting HD do not proceed with testing, mostly because there is no treatment yet anyway.

Another form of genetic mental deterioration is called Tay-Sachs disease (also known as GM2 gangliosidosis or hexosaminidase A deficiency). It is a recessive genetic disorder that starts around six months of age and usually results in death by the age of four. It is caused by a certain allele of the HEXA gene on chromosome 15. A large number of HEXA alleles have been discovered, and new ones are still being reported. These alleles reach significant frequencies in specific populations. French Canadians of southeastern Quebec have a carrier frequency similar to that seen in Ashkenazi Jews, but they carry a different allele. Cajuns of southern Louisiana carry the same allele that is seen most commonly in Ashkenazi Jews.

Since Tay-Sachs disease is a *recessive* genetic disorder, individuals can be *carriers* of the disease (let us give them the simple designation of *Tt*), without having the symptoms. When both parents are carriers, there is a 25% risk of giving birth to an affected child (*tt*) with each pregnancy. The affected child would have received the abnormal allele from both parents. Although there can be several altered alleles in human beings, they all produce a defective protein. The normal protein is a vital hydrolytic enzyme occurring in the lysosomes of a cell—an enzyme that breaks down phospholipids. When this protein is no longer functioning properly, the lipids accumulate in the brain and interfere with normal biological processes.

The consequences are rather serious. Although experimental work is underway, no current medical treatment of the root cause yet exists. Patients receive mainly supportive care to ease the symptoms or extend their lives. Infants are given feeding tubes when they can no longer swallow. Improvements in life-extending care have somewhat lengthened the survival of children with Tay-Sachs disease, but no current therapy is able to reverse or delay the disease's progress; perhaps gene therapy may be possible in the future (see Chapter 9).

Heterozygous carriers (*Tt*) show lower enzyme activity, but manifest no disease symptoms. How come? Enzymes are protein catalysts for chemical reactions; as catalysts, they speed up reactions without being used up in the process, so only small enzyme quantities are required to carry out a reaction. Someone homozygous for the abnormal allele (*tt*) has little or no enzyme activity, so will manifest the abnormal phenotype. A heterozygote (*Tt*), on the other hand, has at least half of the normal enzyme activity level, due to expression by the normal allele. This level is usually enough to enable normal function and thus prevent phenotypic expression.

So far we have discussed two examples of diseases—one dominant and one recessive—that are located on regular chromosomes and follow simple Mendelian rules. There are also genes that are not located on regular chromosomes—also called autosomal chromosomes—but on the X-chromosome. In males (*XY*), the X-chromosome does not have a counterpart, like in females (*XX*), so there is not a matching gene on another chromosome. This makes inheritance for X-linked genes rather different.

Well-known examples of this kind of inheritance are certain forms of muscular dystrophy, hemophilia A and B, and red-green color blindness. Usually such cases are recessive, which means that females possessing one X-linked recessive allele are considered carriers and will generally not manifest clinical symptoms of the disorder. All males possessing an X-linked recessive allele, on the other hand, will be affected, because there is no second X-chromosome that can compensate for this allele. All offspring of a female carrier have a 50% chance of inheriting the mutation. All female children of an affected father will be carriers, for all daughters possess their father's X-chromosome. No male children of an affected father will be affected, as sons do not inherit their father's X-chromosome.

Let us demonstrate this pattern for one particular disease, muscular dystrophy (MD). MD stands for a group of inherited disorders characterized by progressive skeletal muscle weakness, defects in muscle proteins, and the progressive death of muscle cells and tissue. There are nine types of muscular dystrophy, with each type involving an eventual loss of strength, increasing disability, and possible deformity. The various types of MD affect more than 50,000 Americans. The most well-known kind of muscular dystrophy is Duchenne muscular dystrophy (DMD), followed by the much milder form called Becker muscular dystrophy (BMD). The two conditions differ in their severity, age of onset, and rate of progression.

In boys with Duchenne muscular dystrophy, muscle weakness tends to appear in early childhood and worsens rapidly. Affected children may have delayed motor skills, such as sitting, standing, and walking. They are usually wheelchair-dependent by adolescence. The signs and symptoms of Becker muscular dystrophy, on the other hand, are usually milder and more varied. In most cases, muscle weakness becomes apparent later in childhood or in adolescence and worsens at a much slower rate. Fortunately, through advances in medical care, children with muscular dystrophy are living longer than ever before.

Duchenne and Becker muscular dystrophies have similar signs and symptoms and are caused by different alleles of the same gene. The gene associated with this condition is located on the X chromosome. In males (XY), one altered copy of the gene is sufficient to cause the condition. In females (XX), two altered copies are needed to cause the disorder. Because it is unlikely that females will have two altered copies of this gene, males are affected by X-linked recessive disorders much more frequently than females. A characteristic of X-linked inheritance is that fathers cannot pass X-linked traits to their sons.

Usually, an affected male inherits the allele from his mother. A female who carries such a recessive allele is only an unaffected *carrier*. She can pass on the allele but usually does not experience signs and symptoms of the disorder. Occasionally, however, females who do carry one recessive allele may have muscle weakness and cramping too. These symptoms are typically milder than the severe muscle weakness and atrophy seen in affected males.

The DMD gene provides instructions for making a protein called dystrophin. This protein is located primarily in skeletal and cardiac muscles, where it helps stabilize and protect muscle fibers. It is a specific allele of the DMD gene that causes the Duchenne and Becker forms of muscular dystrophy. This allele alters the structure and/or function of dystrophin so it does not work anymore. Muscle cells without enough of the normal protein become damaged as muscles repeatedly contract and relax with use. The damaged fibers weaken and die over time, leading to the muscle weakness and heart problems characteristic of Duchenne and Becker muscular dystrophies. If the abnormal version of dystrophin retains some function, it usually leads to Becker muscular dystrophy. But alleles that prevent the production of any functional dystrophin tend to cause Duchenne muscular dystrophy.

There is one more recent development we need to discuss. It has given a rather unexpected twist to some classical concepts of Mendelian genetics. The idea that genes always occur in two copies (except for the X- and Y-chromosome, of course) and that they harbor dominant or recessive alleles has turned out to be too simplistic. Recent discoveries have revealed that genes can vary in copy-number. Genes that were thought to always occur in two copies per genome have now been found to sometimes be present in one, three, or more than three copies as a result of errors during cell division which can result in duplications or deletions of genes on a chromosomal level.

It had been known for a while already that rare changes in gene numbers were possible; they could actually be detected at the chromosomal level using a microscope. The duplication of the Bar gene in *Drosophila* was one of the earliest structural variations to be linked to a phenotype. Only recently, though, did confusing laboratory results stimulate investigators to ask whether genes are always present in two copies, with a single allele inherited from each parent.

Since then, there have been many findings of variations in copy number. These differences were named *copy number variants* (CNVs). The number of copies can have quite an impact on gene expression. The classical picture tells us that when it comes to genes for enzymes, heterozygotes with an allele for a functional enzyme and one for a non-functional enzyme will not suffer from the non-functional enzyme, because the presence of the other enzyme compensates for it—that is why it is called "dominant." Since enzymes are protein catalysts that speed up reactions without being used up in the process, only small enzyme quantities are required to carry out a reaction. But this picture changes when there can be more or fewer copies of a gene. Gene dosage describes the number of copies of a gene in a cell, and gene expression can be influenced by higher and lower gene dosages.

Here is an example. The human salivary amylase gene (AMY1) is typically present as two diploid copies in chimpanzees. Humans, on the over hand, have an average of more than 6 copies and may have as many as 15. This is thought to be an adaptation to a high-starch diet that improves the ability to digest starchy foods. Approximately one-third of the CNVs observed in the human genome are unique to our species. Another example of a CNV is the copy number of a gene on chromosome 17 encoding for the protein CCL3L1; a higher number has been associated with lower susceptibility to HIV infection. In the same vein, a low copy number of FCGR3B (the CD16 cell surface immunoglobulin receptor) can increase susceptibility to systemic lupus erythematosus and similar inflammatory autoimmune disorders.

Perhaps copy number variation has also something to do with autism, schizophrenia, Alzheimer's disease, and some learning disabilities. There are also indications that certain breast cancers are associated with overexpression of the ERBB2 gene, which codes for human epidermal growth factor receptor 2. Therefore, measuring the ERBB2 copy-number may provide a diagnostic tool for breast cancer and other diseases.

In short, CNVs could very well be more common than previously thought. So the standard model of homo- and hetero-zygotes does not apply to all genes. It is estimated that 12% of the human genome consists of CNVs, and there may be an average of 12 per individual. We will discuss this further in Chapters 10 and 11.

The examples given above are just a small collection of cases showing us the strong impact genes can have on human life. On purpose, I chose very extreme cases that show us the enormous power genes can have in our lives. They were also chosen because they give us an idea of how powerful Mendelian laws and rules have been for genetics and the medical sciences. But the impression I gave you in this chapter is also very one-sided—as if everything in life is determined by our genes. I hope to correct that extreme and erroneous view somewhat in the next chapter.

3. Is It All in the Genes?

The examples of genetic inheritance we have discussed so far have basically a mechanism of a simple type. They are mostly mono-factorial, one-step cases—based on a single gene with one or more alleles that create a specific protein. However, mono-factorial, one-step cases are rather rare.

Simple, although sometimes serious, genetic disorders such as sickle cell anemia do follow the mono-factorial, one-step scenario. It is a genetic blood disorder characterized by red blood cells that assume an abnormal, rigid, sickle shape, which can cause various complications such as anemia. The sickling occurs because of an altered allele for the formation of the oxygen-carrying protein hemoglobin in red blood cells. This is just a simple case of how a gene works—one gene can directly lead to one specific disease if it carries the altered allele. And there are a few more, such as a common form of dwarfism, called achondroplasia, which is caused by an altered allele for a certain cell receptor (FGFR3) that leads to an abnormality of cartilage formation.

Here follows a word of caution first. Even though the previous examples sound very predetermined, they are not the sure thing. Even if geneticists know that everyone with the disease has that one particular altered allele, they do not know whether there are people walking around with that same allele who never developed the disease. The designation "dominant" for an allele should always be a very cautious one.

Next, even if some genes do work in one step, there are many more genes that need several steps to cause differences between people— that is, they require additional steps to take effect. The gene for the ABO blood type, for instance, can hold an allele for antigen A, one for B, or one for neither one of the two (O); these alleles produce an enzyme that then creates the corresponding A- and B-antigens. That is a *two*-step process. Another example would be a single gene that produces a certain enzyme, *tyrosinase*—again a two-step process. Then this enzyme catalyzes the conversion of tyrosine to ultimately the dark skin pigment melanin (via DOPA and DOPA to DOPAquinone). This gene can also harbor one or two abnormal alleles that produce a non-functional enzyme, which then may lead to albinism (but there are also other forms of albinism).

Then there are diseases that require several steps; phenylketonuria is one of them. In this case, there is a gene that produces an enzyme, called *phenylalanine hydroxylase*, but it can also harbor mutated alleles that produce a non-functional enzyme, which then may cause mental retardation, seizures, and other serious medical problems when left untreated with a phenylalanine-free diet. There are several intermediate steps here to lead to this end result.

In addition, there are diseases that require the cooperation of many more genes in order to produce different end-results. The vast majority of human diseases and other genetic traits are in fact multifactorial: They are influenced by many genes interacting with one another as well as by a vast array of signals from the environment of each cell (nutrient supply, hormones, electrical signals from other cells, etc.)—and all these together reflect the external world of the organism as a whole (upbringing, learning, experience, culture, religion). Thus, one and the same change in a specific gene may produce very different results, depending on their surrounding background as well as their genetic background (which represents all other interacting genes taken together), as each human being has a background that is unique. No two persons are completely identical, not even when their genes carry the same alleles, as is usually the case in identical twins. Identical twins do not even have identical finger-prints!

A decade ago, the general estimate for the number of human genes was thought to be well over 100,000, but then turned out to be around 22,500 genes—which is only a little bit more than the 19,735 genes a tiny roundworm needs to manufacture its utter simplicity. And human beings have only 300 genes that are not found in mice. No wonder that the president of Celera, a bio-corporation, said about this surprising finding, "This tells me genes cannot possibly explain all of what makes us what we are." At least, we have a first indication here that genes are not as almighty as some want us to believe.

Not only do genes usually require many more steps to take effect, but they often require the cooperation of many more genes in order to produce specific end-results. These genetic pathways may experience interference from various factors—and these interactions may even be of the nonlinear type. In addition, many genes interact with the environment, so we should rather speak of a genetic "predisposition" than of a genetic determination. A genetic predisposition means that an individual has a genetic susceptibility to developing a certain trait, but is not destined to develop that trait.

Actually, the development of many traits depends in large part on a person's environment and lifestyle. A simple example of the interaction between genetic predisposition and environment is the higher susceptibility of fair-skinned people to the development of skin cancer. In these lighter-skinned individuals, skin cancer can be associated with changes in the melanocortin-1-receptor-gene (MC1R), but is also dependent on the length of exposure to sunlight. Another example of a genetic link to an environmental factor is seen among people with a deficiency of alcohol dehydrogenase (ADH) caused by a change in the ADH1C gene, which encodes an ADH subunit. Only when they drink alcohol, even in small amounts, ADH deficiency can become manifest.

There are many other complex traits that arise from numerous genetic and environmental factors working together. A good example is human height: It has a heritability of 89% in the United States, but in Nigeria, where people experience a more variable access to good nutrition and health care, height has a heritability of only 62%. Apparently, genetic simplicity has given way to complexity, and hence to a more complex kind of genetic determinism—but it still is determinism in essence, although not of a simple kind, let alone of the mono-factorial type.

Yet there are still some scientists who like to think in mono-factorial terms. In their view, the search is on for that one specific gene that determines that one specific character in one's life. Someday, they hope to discover, for instance, what that one fundamental genetic cause is behind all cases of autism. However, the range of symptoms is so varied that it probably is a catch-all term to describe a collection of related disorders. The genetic research on it is quite good, but also very complex implicating numerous variants in a variety of genes. Vitamin D deficiency may be one of many possible environmental triggers. Clearly much research remains to be done. In science, discoveries always start as inventions—usually called hypotheses.

However, not all inventions lead to discoveries. To use an analogy, the person who invented "Atlantis" did not discover Atlantis; it remains a legendary island until further notice. The same in science: Most inventions do not lead to discoveries. Yet some scientists think they have made a discovery when all they have in mind is an invention, a hypothesis. As a consequence, we have been bombarded with new genes: a gene for longevity, a gene for homosexuality, a gene for schizophrenia, even a gene for religion—the list could go on and on.

I am not claiming here that genes can affect behavior. After decades of research, it is widely accepted that most behaviors in animals and humans are under some degree of genetic influence. Model genetic species, such as *Drosophila melanogaster* (common fruit fly) and *Apis mellifera* (honey bee), have been proven to be instrumental in developing the science of behavioral genetics. It turned out that the gene regulating foraging behavior in fruit flies also regulated social behavior in honey bees. A so-called "foraging gene" shows allelic variation that produces behavioral syndromes in fruit flies, from "sitters" to "rovers". Honey bees have a variant of the same foraging gene that controls the onset of foraging behavior.

Perhaps this kind of research can also be helpful when it comes to human behavioral traits such as schizophrenia. Using known functions of brain systems and neurotransmitter systems, scientists correlate behavior to these brain areas (e.g., excess glutamate release may stimulate excess dopamine in the limbic system leading to schizophrenic symptoms). Once scientists are able to map behavior to biological systems, they can then turn to genetics to understand the development of these biological systems (e.g., an abnormal glutamate gene could be a candidate gene for schizophrenia). But again, genes never work in isolation, let alone in a vacuum.

Reacting to the casual way in which some neurobiologists speak of a gene for depression or a gene for violence, one could argue that depression and violence are only labels for rather complicated and variable patterns of behavior. The hypothetical genes some have come up with were once claimed, and then had to be retracted; they were often inventions that did not lead to discoveries. They were based on the mantra of mono-factorial determinism, but that form of determinism only exists in the setting of simplified models (more on models in Chapter 5). Nevertheless, genetic determinism is still rampant in science, hence hypothetical genes just keep coming and going. I would say that is where science borders science fiction. I could even come up with a gene that makes one believe in the all-powerfulness of genetics!

I am not declaring all these inventions bogus, but most of them are still in the stage of invention and are awaiting the stage of discovery. And yet it remains tempting to claim a genetic difference for something that may not even exist. Perhaps alcoholism is not genetic but rather something acquired at home or in the womb or in a group of peers. Perhaps pedophilia is not an issue of genetics but rather a form of immoral behavior—rape, that is. Perhaps Munchausen Syndrome is only a call for attention or sympathy. I am certainly not trying to put all these examples in the same basket, but I just want to signal how often we get bombarded with a new "disease." First they invent a disease, then they invent the gene to explain such a disease—and then they sell us an elixir to cure it! But that is what late-night infomercials want us to believe, in denial of the fact extensive clinical trials must be conducted for release of new pharmaceuticals. During this process it may turn out that we may have been dealing with phantom diseases and bogus genes.

Let me illustrate the many pitfalls this kind of research could be in for with the following test case. The American human geneticist Dean Hamer and his team once postulated a "gay gene." First I must acknowledge that Hamer did not claim that the gene itself causes male homosexuality, but rather its variants that may have an influence on phenotypic variation. He also was aware of the fact that this was not *the* gay gene but rather one of potentially many whose variants might influence predisposition to male homosexuality. But most of those caveats did not reach the mass media.

The first problem of Hamer's team was to localize this hypothetical gene locus on one of the 23 chromosomes. They found a possible candidate when they asked 76 gay men about their relatives. It turned out that male homosexuality occurred more often among uncles and cousins on the maternal side than on the paternal side. All 655 relatives were asked to answer Kinsey's questions about sexuality. Although only 30 lists were returned, there was a significant difference between the maternal and paternal side. Apparently—assuming both sides answered with equal honesty—the hypothetical gene is located on a chromosome that sons receive from their mothers, not their fathers.

So the search was focusing in on the X-chromosome.

Then they used so-called "genetic markers" on the X-chromosome—genes for hemophilia and color blindness, for instance, that had a known position on that chromosome. We know from the first chapter that the chance of recombination between two genes is proportional to the distance between those two genes. In the study under investigation, recombination can only be recorded if we have at least two gay brothers in each family and preferably also their mother. This requirement was only fulfilled in 40 families (but, for some reason, only for 16 mothers). If two gay brothers share not only their assumed gene for homosexuality but also a particular genetic marker, then that marker must be close to the gene we are trying to locate. Well, this happened to be the case for five markers at the end of the long arm of the X-chromosome. From the 40 pairs of brothers, 33 pairs had those five markers in common (13 more than the 20 to be expected by mere chance).

It is clear in this case that we are working here with Morgan's definition of a gene: A gene is the smallest unit of recombination. Did Hamer a.o. really find and locate a gene for homosexuality? That remains to be seen. Not all inventions are discoveries, we said earlier. In fact, their research is based on many presuppositions that need further scrutiny.

First, based on Hamer's research, all we know is that there might be some gene involved, but we do not know what came from other genes and the environment. To find out how strong a gene's impact is, we would need research on adopted identical twins as well (more on this in Chapter 4). We do know from a 1991 study of identical twins that 52% (29 pairs out of 56) of the identical twins were both homosexual; 22% (12 pairs out of 54) of the fraternal twins were both homosexual; and 11% (6 of 57) of the adoptive brothers where both homosexual. If homosexuality is genetically determined, why did only 52% of the identical twins share the same sexual orientation? How about the other 48% of the twins who differed in their sexual orientation? These are serious questions that need to be addressed, and have been more or less in later research

Second, Hamer's team had preferentially recruited twins where both brothers were gay by advertising in homosexual newspapers and magazines rather than in periodicals intended for the general public. Indeed, in a more recent 2000 study, the same researchers used volunteers recruited, not from the gay community, but from the Australian Twin Registry. They found that only 20%, and not 52%, of identical twins share the same homosexual orientation. Obviously, it is very difficult to distinguish the genetic from the environmental influences on sexual orientation. In short, genes may have some impact but certainly not a monopoly.

Third, Hamer's research had a strong statistical basis. Even if their findings are deemed statistically significant, that is always based on a certain confidence level—usually of 95 percent. A 95% confidence level implies a 5% error chance of accepting a spurious significance, which is called an α-error or Type I error in statistics. Besides, statistical data can only detect associations, but not causality per se. Does the gene under consideration really cause homosexuality, or does it cause something that is related to it? Reasoning in terms of *Post hoc ergo propter hoc* (Latin for "after this, therefore because of this") is a logical fallacy that erroneously states, "Since some event was followed by another event, the latter event must have been caused by the former one." We know this is certainly not always the case: When the barometer drops, we can expect rain, but the barometer does not cause the rain—it is the change in atmospheric pressure that does the trick for both of them!

Perhaps, in this line of thought, the hypothetical gene on the X-chromosome is not a "gay gene," but affects excessive maternal care instead. Did Hamer not say that mothers give this gene to their homo-sexual sons? Well, could it be that mothers with this gene "create" homosexuals? Psychoanalysts might tend to think so. I am not saying they are right, but I do say that there is no very strong evidence here at all. All too often similar hypothetical genes—for example, genes for schizophrenia, autism, and bipolar disorder—were claimed, and then had to be retracted. They were inventions that did not lead to discov-eries. Hypothetical genes just come and go in genetics. Admittedly, Hamer never claimed that there was a single X-linked gene that causes male homosexuality.

Fourth, Hamer's discovery was followed by further scientific research that sometimes contradicted his findings. In fact, two subsequent, much larger studies of other homosexual brothers—one by J.M. Bailey a.o. and one by G. Rice a.o.—have concluded that there is no evidence that male sexual orientation is influenced by an X-linked gene. Rice's data did not support the presence of a gene of large effect influencing sexual orientation at position Xq28. However, meta-analysis of all data available has indicated Xq28 may have a significant but not exclusive effect. A subsequent genome-wide scan by Hamer's group revealed several additional regions on autosomes that were moderately linked to male sexual orientation.

Fifth, we would also need to come up with a mechanism that makes it possible for an inherited tendency for homosexuality, if there is such a thing, to persist in populations. An overly simplistic model would sug-gest that if homosexuality were strongly genetic, natural selection would weed it out—the same way it would eliminate celibacy if that were based on genes, since such genes could never reach the next generation and would therefore rapidly decline in frequency. In more complex selection models, however, a higher fecundity among the female relatives of homosexual or celibate men is one of several ex-planations for the inability of natural selection to remove alleles that may cause such genetic predispositions. But the question remains whether this can also explain that homosexuality currently seems to be on the rise.

Sixth, homosexuality would be many steps away from a specific gene. Hamer was very well aware of the fact that a single genetic locus does not account for all of the observed variability. As we know nowadays, the absence or presence of a Y-chromosome does lead to a sexual difference; it is more in specific the SRY gene on the Y-chromosome that acts like a master switch and is responsible for the development of a fetus into a male by initiating the testes development. However, we should not confuse a difference in sex with a difference in *gender*. Homosexuality is not a matter of sex (of producing either ova or sperm), but rather of gender, which suggests it is many steps away from the genotype, and hence, allows for many inroads from the environment.

A difference in gender entails much more than a difference in biological characteristics—namely also differences in behavioral traits, social roles, and cultural expectations that come with being a man or a woman. Early on in human development, parents as well as society take on a molding role. As soon as parents know their child is a boy, they treat it as a boy, which makes the child consider himself as of the male gender. So the question is whether differences in gender are only the outcome of differences in sex, which in turn are supposedly based on differences in genes—or is there perhaps much more going on?

Apparently, the distance between genotype and phenotype is a very long one here. There are also sex hormones involved; however, genes do not produce sex hormones, but rather the enzymes that in turn produce hormones—which is at least a two-step process. Even if you would argue that genes affect hormones, and that hormones affect the brain, and that the brain affects our behavior, then we should emphasize that this behavior, in turn, affects the brain again. A similar phenomenon is well known from sports, for instance: Strong muscles benefit those who play sports, but in turn, playing sports greatly benefits the development of the muscles.

In other words, it is not only the genes, the hormones, and the brains that shape our behavior, but everything that we see and hear around us, plus all the dreams, ideals, hopes, plans, and expectations we foster in our minds. All these have an impact on the way we develop ourselves. What seems to be "in-born" may in fact very well be "in-printed" or even "self-taught." If we have children, we are supposed to *raise* them—which is more than giving them our DNA. That is where nature and nurture meet. However, very often, or maybe even always, it is nearly impossible to tell what part is "nature" and what part is "nurture." And yet, most of the time, there is also a nurture part—for the simple reason that it is not all in the genes. This takes us to the next chapter.

4. Nature versus Nurture

When I said that what seems to be "in-born" may in fact very well be "in-printed" or even "self-taught," I was actually looking for a way to peel "nature" and "nurture" apart. Is that possible? Some claim they have found such a way. The best, and probably only, way is a combination of adoption research and twin research—that is, identical twins split by adoption. Let us find out why.

Adoption research, on its own, is not good enough. Although it is clear that relatives resemble each other more than strangers—because they have more alleles in common—parents do not only give alleles to their children but also parts of their environment. Even adopted children did share, for some nine months, the surroundings of their mother's womb, including her voice, her hormones, her food, her moods, and even her medications. Plus adoption usually takes place in an environment that is very similar to the original one, often just around the block or with relatives or friends. So, we tend to easily over-estimate the impact genes have if we go by adoption research alone.

On the other hand, twin research, on its own, is not good enough either. Identical twins have, in general, the same alleles, but as we discussed already, they can vary in the number of copies they have of certain genes (CNVs). New CNVs have been found between identical twins who otherwise have identical genomes. In addition, similarities between identical twins are not only the result of identical alleles but also of almost identical surroundings—the same womb, that is, often even the same placenta. Besides, their strong resemblances make it more likely that they will be treated by others the same way later on in life. Moreover, they often strongly desire to become and be more like each other. So also in this case, we would over-estimate the impact of genes.

As a consequence, whether it is twin research or adoption research, we tend to underestimate the "nurture" component. Another complicated factor is the fact that twin and adoption studies tend to produce different estimates of the genetic component. Combining them is perhaps the best way we have to correct for the fact that they are rather skewed on their own.

Another problem with this kind of research is that we need samples that are not biased but represent the total population accurately. I remember a case where scientists studied members of an organization of male homosexuals and found that these people came from families with a significantly higher percentage of male siblings—whatever that means. What they did not seem to realize, though, is that families with only female offspring were not represented in their sample at all, and thus they were *under*-represented in proportion to the percentage of female siblings in other families. To rectify this omission in a statistically correct manner is extremely complicated. Not all scientists, not even geneticists, are expert statisticians.

Statistical research stands or falls anyway with the size of the samples used and the competence and discipline of its users. It should alarm us that on March 29, 2012 the high-standard scientific journal *Nature* published a disturbing commentary claiming that in the area of preclinical research—which involves experiments done on rodents or cells in petri dishes, with the goal of identifying possible targets for new treatments in people—independent researchers doing the same experiment could not get the same results as reported in the scientific literature. Over 10 years, Amgen researchers could reproduce the results from only 6 out of 53 landmark papers. And researchers at Bayer Healthcare reported that only in 20-25% of 67 projects analyzed the relevant published data were completely in line with their in-house findings. I consider that alarming news.

In addition, let us never forget that, along with differences in genetics, there are also personal and cultural variations in most human characteristics—another strong nurture component. We often follow habits acquired at home, in school, through peers and friends, and through the society we live in—but that does not mean they are *genetic*. The fact that being a Democrat versus a Republican or being a Christian versus a Muslim or being a Red Sox fan versus as Yankee fan runs in families does not mean it is a matter of genes. To put it differently, there certainly is not a gene for everything. If you never touch alcohol or illegal drugs, you will never develop an addiction for these substances. If you did touch them, though, would that be a genetic compulsion or rather peer pressure, or an act of depression, or even mere curiosity? Probably any or all of the above!

Of course, we could speak of a genetic "predisposition" instead, as I did before, but that is a very vague notion—and often misleading besides. People who "tend to be fat" on 5,500 calories a day "tend to be thin" on 2,000. To say that "X has a predisposition to be fat" might simply mean that, on 4,000 calories a day, X is fatter than the average person on a comparable diet. In a similar vein, those who are supposed to have a "predisposition for addictions" still need the "proper" environment for those addictions to materialize. And besides, even these so-called predispositions can yet be reprogrammed (more on addictions in Chapter 13).

In short, it is hard to believe there is such a thing as a "chip gene" for people with an addiction to chips, a "chocolate gene" for chocoholics, or a "spending gene" for habitual big spenders. If "health anxiety" (hypochondriasis) runs in the family, it could very well be that an overly protective caregiver or an excessive focus on minor health concerns in the home is to blame. I could even think of a gene for playing ball games—with an allele for soccer, predominantly present in South-Americans, and an allele for football, exclusively existent in North-Americans. I don't think that makes for a strong case.

Genetic determinism certainly has its limitations. Just think of the fact that we could never compute an organism based on its genes since an organism does not compute itself from its genes. We are not at the mercy of our genes and our genes are not our destiny; they are like a hand of cards we are dealt, but we can play them differently. Human behavior may be more often than not a matter of lifestyle *choices* rather than the outcome of a set of genetic instructions. Even if there are genes that determine a certain behavior, we need to realize that genes often must be "turned on" by an outside force before they can do their so-called "preprogrammed" job.

The outside force of high levels of stress, for instance, may activate a variety of genes, including those suspected of being involved in fear, shyness, and even some mental illnesses. Children conceived during a three-month famine in the Netherlands during a Nazi blockade in 1945 were later found to have twice the rate of schizophrenia as did Dutch children born to parents who were spared the trauma of famine. Behavioral epigenetics attempts to provide a framework for understanding how the expression of genes is influenced by experiences and the environment to produce individual differences in behavior.

Perhaps this is the right moment to make an important distinction between the words "hereditary" and "genetic." Heredity is passing of traits to offspring, but this can be a matter of either genetics or upbringing. Some traits may be hereditary but not genetic—probably humor, religion, and the like. Owning lots of money may be hereditary but certainly not genetic. Some other traits, on the other hand, may be genetic but not hereditary, because they are the result of genetic changes, called *mutations* (see Chapter 10), occurring in a specific cell line of the organism (unless the change occurred in the cell line of reproductive cells). People with two different eye colors or with a patch of white hair probably had a mutation in certain cell lines, so they will not pass this on to the next generation—it is a genetic but not hereditary feature.

Perhaps another example might help me make my point even more clearly. There is a gene which codes for an enzyme called mono-amine oxidase (MAO). When it works correctly, MAO breaks down certain neurotransmitters such as serotonin, dopamine, and noradrenaline. When it does not, these neurotransmitters build up. For example, one common, low-activity allele (MAOL) was found to be linked to aggression and gang membership in some boys. Another study found that MAOL was linked to antisocial behavior, but only in boys who had an abusive childhood. The high activity version, MAOH, was linked to fraud, but only in people associated with other delinquents.

In other words, this gene does not harbor alleles that directly *cause* aggression. The effect it has on behavior depends also on the environment—that is, nature *and* nurture. But in fact the case is even more complicated than that. One study on pregnant women found they were more likely to become depressed if they had the MAOL allele. But it turned out that they were only more likely to become depressed if they were also carriers of another gene (COMT).

Apparently, the effect of a gene and its alleles varies greatly depending on which other genes they have pitched up with in the particular individual's genome. The trait a gene is linked to changes depending on which other genes a person carries. As a consequence, it is extremely difficult to link genes to personality—so difficult, in fact, that no one has done it yet.

It is not just in psychology that we find this problem. We find something similar for adult-onset (type-2) diabetes. Diabetes is a very complicated disease, with a huge number of genes involved, all of which interact in complex ways with each other and with the wider environment. There is just no single "gene for adult-onset diabetes," but there are genes that, in the presence of certain other genes, and in the "right" environmental context, seem to increase someone's risk for type-2 diabetes. Although this kind of diabetes runs in families, this tendency is also partly due to children learning bad habits of eating a poor diet and not exercising—partially learned from their parents.

Perhaps we can rephrase the problem we have here in terms of the following question: If there are genes for eye-color, color-blindness, etc., would the sum of all these together be enough for having eyes? Think about that question for a moment. Probably the best answer to this question is that a genotype is more than a genome. A genome is just a cluster of genes, but a genotype is the blueprint of a genome plus the instructions for processing the blueprint. In other words, a gene does what the program dictates; but it is not the other way around—the program does *not* do what a gene dictates. A genotype is more than a bunch of genes.

Unfortunately, the "genetics of the genotype"—some call this epigenetics—is currently not as well developed as the genetics of the genome. Our genome acts more like an archive with instructions, but *what* to use from this archive *when*, *where*, and *how* is partly determined by the genotype, but definitely in very intricate ways.

Usually it is assumed in the nature-nurture debate that it all starts with "nature," but that "nurture" may interfere during the rest of the process. Genetics assumes that there is a *one*-directional flow from genotype to phenotype—in other words, the genotype can change the phenotype, but the phenotype cannot change the genotype. While it is possible for genes to change someone's behavior, there is no evidence to suggest that one's behavior can change the genes. Bodybuilders do not produce descendants who have bodybuilder features already built into them.

In general the previous assumption may be true, but as we saw already, sometimes "nurture" is in fact able to switch certain genes on or off. Sure, this does not alter the fact that the genes themselves do not change; even if they do change, that is not in response to a certain "need" triggered by "nurture" (see Chapter 9). Yet, there are some indications that environmental influences can etch chemical modifications in DNA. Subtle chemical markings on the DNA of stem cells recovered from the umbilical cord blood of babies, for instance, turned out to be significantly different for different body sizes; so there might be influences in the uterus that do affect genetic mechanisms. Epigenetics may be on its way to become a hot new field of biology.

Epigenetics is the study of heritable changes in gene activity which are not caused by changes in the DNA sequence. Examples of mechanisms that produce such changes are DNA methylation and histone modification, each of which alters how genes are expressed without altering the underlying DNA sequence (see Chapter 12). DNA methylation turns a gene "off," resulting in the inability of genetic information to be read from DNA; so removing the methyl tag can turn the gene back "on."

Later on we will see that the idea of a one-directional flow from genotype to phenotype translates into the one-directional flow from DNA to mRNA to proteins (see Chapter 9). There does not seem to be a way back from altering proteins to altering DNA. "Nurture" cannot change "nature"—as far as we know. But as with everything in biology, there are always exceptions. A retrovirus, for instance, is an RNA virus that is duplicated in a host cell using the reverse transcriptase enzyme to produce DNA from its RNA genome. This is actually a case of transcription the other way around—the transcription of RNA from DNA in reverse. Even human cells have this capability. Self-replicating stretches in the human genome known as retro-transposons utilize reverse transcriptase to move from one position in the genome to another via an RNA intermediate. Telomerase is another reverse transcriptase found in humans, which carries its own RNA template; this RNA is used as a template for DNA replication. Because this idea goes against an old dogma of genetics—a one-directional flow of genetic information—it may not have received as much attention in genetic research as it deserves (more on this in Chapter 11).

Let us come to a conclusion. Human diversity is immense. Some are handy, some are clumsy; some are aggressive, some are docile; some are neat, some are messy; some are introverts, some are extroverts; some are go-getters, some are laid-back; some are easy-going, some are strict; some are insensitive, some are compassionate; some are humorous, some are serious—the list could go on and on. It is often far from clear how these people became that way; perhaps they were "born" that way, perhaps they were "made" that way, or maybe they "chose" to be that way.

If teachers would tell their students that genetic research has shown that the allele for a blue eye color also positively affects intelligence, you can almost bank on it that blue-eyed students will soon perform better in class.

The bottom line is that there is always more to it than merely genes. Even most diseases nowadays are not a matter of genetics but a result of our uncontrolled appetites. Genes may be everywhere, but genes are not all there is. Let me repeat it again: It is just not all in the genes!

II. The Battle of Genes

5. The Gene Pool

We discovered in Chapter 1 that one of the successful moves Mendel made was that he did not attempt to study everything about all the offspring at once. Instead, he selected only a few individual traits of pea plants and studied them in detail. Mendel knew how to select and isolate things by limiting himself to a small selection of specific details.

What Mendel did is something all good scientists do or should do. Good scientists are those able to demarcate their area of investigation, by limiting themselves to factors that are relevant to what they are studying, and keeping strict control over factors that might interfere with their search. To put it more technically, scientists create a simplified *model* of what they are studying. They reduce complex entities to a manageable model related to an analyzable problem as a successful way of doing research. As Peter Medawar used to say, science is "the art of the soluble."

A model is like a simplified replica of the original—that is, with a controlled setting and a limited scope. Models are handy tools to study things in a "test-tube-like" setting, removed from the complexity of the real world. It is in the "test-tube" of a model that all boundary conditions can be isolated, while other interfering factors can be kept under control.

That which has made some scientists so successful is the fact that they were able to create a test-tube-like shelter in a laboratory removed from the complexity of nature so that the various factors under investigation can be isolated and manipulated on an individual basis. Science works with models—and models always select, so they neglect what they did not select. Apparently, science purchases success at the cost of limiting its ambition. Models can certainly enhance our view— with a better focus, a higher precision, and more details—but at the same time they can also narrow our view by obscuring what is outside their scope. That is the good news and the bad news.

Take the case of a gene-pool model—a model very popular in popula- tion genetics. The gene pool represents the set of all genes in a popu- lation. The simplest models of a gene pool focus on one specific gene with its various alleles; they simulate how selection pressure on this particular gene may change allele frequencies by selectively promot- ing certain alleles of this gene. So the simplest models simulate a pro- cess of natural selection among the different alleles of one single gene in a population (see Chapter 6). To put it briefly, it simulates the battle of genes—or in simple models more specifically the battle of alleles based on one particular gene.

Let me explain this concept in a more technical way. Mendel's first "law" describes how genes are passed from one generation to anoth- er. If the gene pool is composed of two alleles, say A and a, and the frequency of allele A is p, then the frequency of allele a is $1-p = q$. From this, the frequency of the different genotypes can be deduced as follows: The frequency of genotype AA is $p \times p = p^2$, the frequency of genotype aa is q^2, and the frequency of Aa plus aA is $pq + pq = 2pq$.

Further mathematical analysis shows that if all genotypes mate at random with one another, then their frequencies in the next genera- tion will remain stable. Conclusion: If the frequency of AA was p^2 in the previous generation, it will remain p^2 in the generation(s) to come. Similar calculations show that the same holds for the other genotypes as well. It is the gene-pool model that makes this approach possible and fruitful.

In case you are interested, this is called the Hardy-Weinberg law (or theorem), which states that allelic frequencies and genotypic ratios remain constant from generation to generation—under the condition, of course, that reproduction is random and disturbing factors are absent. However, in real life, that is seldom the case! First of all, alleles may undergo a "random walk" process, called genetic drift. Second, certain alleles have a better chance to reach the next generation than others do. The "abnormal" alleles we talked about in the previous paragraphs may not have as much chance as the "normal" alleles to get transferred to the new generation. In other words, their "fitness" has been reduced; they have a selective disadvantage, which is called a negative selection pressure. As a result, the frequencies of alleles may very well change in future generations instead of staying the same.

How can alleles differ in "fitness"? Research has shown that most populations—gene pools, if you will—harbor a tremendous amount of variability. How is that possible? Would the "normal" allele not gradually outpace all "abnormal" alleles? To be sure, the "classical" school used to claim that each allele may have its own selective value, with one of them considered the "best"—which they often called the "wild type" (and I did something similar before by speaking of "normal" alleles versus "abnormal" ones). The "balance" school, in contrast, denied that one specific allele is the absolute "winner," because certain alleles may *together* be the "best" fit. So it may not be quite appropriate to speak of "normal" and "wild-type" alleles.

How can we simulate this phenomenon of changing allele frequencies within a simple gene-pool model? Here is the answer. By introducing selection ratios into the previously used formula, shifts in allele frequencies can actually be calculated and predicted. Take, for example, the well-known case of sickle cell anemia—which we discussed earlier. The sickling occurs because of the presence of an "abnormal" allele for the formation of the oxygen-carrying protein hemoglobin in red blood cells. It is caused by a recessive allele—let us label it a (there are other abnormal variants as well).

Because there is strong selection pressure (s) against the homozygote (aa), who suffers from anemia, we would expect this allele (a) to gradually disappear from the population. However, in malaria areas it has a rather stable frequency (q). The explanation is that there is also selection pressure (t) against the other homozygote (AA), who is more vulnerable to malaria than the two other genotypes. It turned out that, in areas where malaria occurs, the heterozygote (Aa) has a higher level of fitness to resist both malaria and anemia, and thus keeps the recessive allele a "alive" in the population.

With the help of some mathematical manipulations, we are able to deduce that the frequency of alleles will become stable as soon as $tp = sq$. This is a form of balanced polymorphism—an equilibrium mixture of homozygotes and heterozygotes maintained by some negative selection against both homozygotes. So the conclusion is that allele A is not necessarily the best fit in just any kind of environment.

This explains why the sickle-cell allele can remain present in certain populations—in spite of its damaging anemic effects. As a matter of fact, the sickle-cell anemia is not uncommon in parts of tropical and sub-tropical regions where malaria was or is still common. The highest frequency of the allele is found in northern Angola (at $q = 0.1818$, with sickle-cell anemia at $q^2 = 0.03$). Frequencies like this can be explained by the fact that in areas where malaria is common, there is a positive survival value in carrying just one sickle-cell allele.

This explains why Americans with ancestors from these regions still have a relatively high frequency of the sickle-cell allele. The prevalence of sickle-cell anemia is 1 per 500 among African Americans, and 1 per 36,000 among Hispanic Americans. The average life expectancy of persons with this condition was estimated to be 42 years in males and 48 years in females, but today, thanks to better management of the disease, patients can live into their 50s or beyond. For the record, it is estimated that 2.5 million Americans are still heterozygous carriers for the sickle cell trait.

It is the gene-pool model that made the previous analysis possible. However, not only can the gene-pool model be used for a single gene, but also for a series of genes, or even the entire set of genes in a population. Some people call this kind of approach disrespectfully "bean-bag genetics"—a bag of genes in a bag of chromosomes in a bag of cells. Yet, this approach can be a very sound scientific strategy—sound until it goes astray. It is a sound technique until we mistake the gene-pool model for what it represents. The danger is that the model may give us the impression that the population is actually nothing but a set of genes—or more specifically, that the units of selection are in fact genes and their alleles, but that is not so "in the real world" where the organism, not the gene, is the unit of selection.

Why are alleles not the "real" units of selection? The answer is that selection simply cannot "see" certain alleles, because recessive alleles—which do not come to expression in the organism—cannot be subject to selection, since selection just cannot "detect" them. Alleles may be considered the units of selection, but only so in the model, not in the real world. In addition, selection may not even "see" individual genes, but only clusters of genes that have an impact on each other.

Because of all of this, we should be on the alert not to confuse models with the reality they refer to—otherwise we might end up with all sorts of scientific nonsense such as claiming that alleles are "selfish" elements engaged in a constant "survival battle" with other alleles. Take it as playful metaphorical language, but nothing more. Or consider this one: All organisms are essentially mere "survival machines" for their genes. Each time, the deceiving keyword here is hidden in the term "essentially"—suggesting that the model is equivalent to the real world, whereas it is just a replica simplified for scientific purposes. This kind of talk is no longer scientific talk, but science-fiction talk at best.

Does this potential glitch make the model useless? Far from that, but do not let this lure you into identifying the real world with one of its models—that would be an illegitimate move. A limited technique of simplifying what is complex does not entitle us to declare all complexity as a matter of mere simplicity! Those who say that a population is just a collection of genes, or alleles, are actually confusing an abstract model with the real world it is a simplified replica of. Be aware, the model is just a scientific tool, not a new world-view. If we lose that awareness, science will easily slide into an ideology or world-view.

I think I have stressed enough that science becomes arrogant as soon as it oversteps its limits. All models are by nature an abstract replica of what they represent. So let us not fall for the timeless temptation to simplify the vast complexity of reality by substituting reality with one of its simplified models. Do not declare the simplified model to be the new reality—not even the gene-pool model. If you decide otherwise, you would be ripped off with a stripped version of reality, because a model is only a surrogate for "the real thing." The only model that could ever qualify as a perfect replica of the original is the original itself—but that would not be a model anymore.

Let us go back to the gene-pool model to answer one more question: Can the gene pool—being the set of all genes in a population—ever become split, so that one part gets isolated from the other part? Why would this be an important issue? The answer is that people who believe in evolution claim that the gene pool of a species can be split into two isolated gene pools, thus creating two different species. Was that not Mendel's main question?

Species are defined as groups of interbreeding natural populations that are reproductively isolated from other such groups. This means, in terms of a gene pool, that the two pools have been isolated from each other and are no longer able to share in each other's genes and their alleles. Gene exchange is prevented by a "barrier." The barrier between the two gene pools is very often of a geographical nature—physical barriers such as mountains, deserts, or water—but it can also be the mere physical distance between some members of a population. Because of such barriers, the two gene pools develop their own characteristics and may no longer be able to exchange genes and alleles. So the physical barrier becomes gradually a biological barrier as well. Biologists have found numerous cases of such a situation.

What about hybrids, you might wonder. Hybrids seem to be an exception to the rule, but they are arguably not. Although hybrids are the offspring of two organisms belonging to two different gene pools, the separation between the two pools is still in the process of being developed but has not been finalized yet. Hybridization may still occur after two populations of the same species have become more or less separated but then come back into contact with each other before separation has completely set in. Sometimes, the separation of gene pools can still be at an early stage, such as with crossings between the Bengal tiger and Siberian tiger; sometimes, the process seems to be farther ahead, such as with crossings between lions and tigers; very often, the separation has become extensive, such as with crossings between sheep and goats, so hybrid embryos usually die before being born.

Consider the well-known case of a crossing between a horse and a donkey. A horse has 32 pairs of chromosomes, while a donkey has only 31 pairs. Hybrids are called either a mule, when the father is a donkey, or a hinny, when the mother is a donkey. Most hybrids are infertile because the chromosomes have to match up properly to create viable embryos. However, female hybrids (both mule and hinny) can sometimes still produce offspring, whereas all male hybrids are infertile (probably because they miss a 2nd X-chromosome).

It might be interesting to bring Mendel into this discussion. When Mendel did his "hybridization" experiments, he recognized that the line between varieties and species was often blurred. In his own words, "It has so far been found to be just as impossible to draw a sharp line between the hybrids of species and varieties as between species and varieties themselves." To put things in our current terminology, was he dealing with members of two different species—gene pools according to our modern terminology—or were they actually from the same gene pool.

Here are some examples of gene pools that are somewhere in the process of being split into two isolated gene pools.

1.   The tormentil shrub of the coastal area in California is pretty large, whereas the tormentil shrub of the Californian mountain regions appears stunted. They can no longer interbreed because transplanting them to the other region causes them to die. This may be more of a borderline case of splitting the gene pool; nonetheless, the outcome is that these two populations are in fact reproductively isolated from each other. They could very well be on their way to becoming two different species, for a species is defined as a reproductively isolated population—or gene pool if you will.

2.   The locusts *Chorthippus bruneus* and *Chorthippus bigutullus* can no longer interbreed because they use a different mating call. However, they do still interbreed when the mating call of their own species is played on a recorder. But some day, this may no longer be possible because additional differences may have accumulated as well, causing them to be completely isolated from each other in their reproduction. Their gene pool has been virtually split.

3.   The bishops-pine *Pinus muricata* and the Monterey pine *Pinus radiata* cannot pollinate each other in nature since they have developed different pollination times, but they can still do so with manual intervention. The gene pool split may not be final yet, but it seems to be on its way.

4.   The African Elephant and the smaller Asian elephant are definitely two different species. There is much evidence, however, that within the African species new internal reproductive splits have developed due to geographical isolation, and therefore, the elephants of West Africa may have to be regarded as a separate species from either the savanna elephants of Central, Eastern, and Southern Africa, or the forest elephants of Central Africa.

5.   Domestic animals can usually still produce fertile offspring with the original members of their species —probably because the split was rather recent on an evolutionary scale. But, there are some exceptions. Domestic sheep, for example, no longer produce viable offspring with *Ovis orientalis*, one species from which they supposedly descended. On the other hand, crosses between sheep and goats are still possible, but their embryos die early in development. So in effect, they are reproductively isolated as well, due to a stunted embryological development of hybrids.

6.   Another example would be the two species of gulls that are connected with each other through a continuous ring of subgroups. The black-backed gull *Larus fuscus* probably started in Asia, changed westward into a Scandinavian and a British subgroup, and eastward into three other subgroups—one in Siberia, one in America, and another one back in England. Each subgroup will interbreed with the adjacent one in the ring, but the two subgroups at the end of the ring— that is in England—can no longer interbreed, because their genetic differences have accumulated over time and distance. The British herring gull *Larus argentatus* will not interbreed with the British black-backed gull *Larus fuscus* under any conditions. The first one nests on cliffs and does not migrate; the second one breeds inland and migrates for the winter. They no longer belong to the same gene pool.

7. A similar case can be found in California, where the *Ensatina* salamander forms a horseshoe shape of populations in the mountains sur-

rounding the Californian Central Valley. Although interbreeding can happen between each of the nineteen populations around the horse-shoe, the salamanders on the western end of the horse-shoe cannot—or can no longer, I should say—interbreed with the salamanders on the eastern end. The question, then, arises whether to classify the whole ring as a single species (in spite of the fact that not all individuals can interbreed) or to consider each population as a separate species (in spite of the fact that there is still interbreeding with nearby populations). This much is clear: If enough of the connecting populations within the ring perish to sever the breeding connection, the remaining members would have become two distinct species.

It should not surprise us then that, in an evolutionary context, a clear-cut species concept inevitably becomes a bit fluid. The situation is again comparable to what happened to English in the USA. People from the North may have some difficulty understanding people from the South, since they have developed their own dialects being so far away from each other. But they have not made it (yet) to different languages. It is a poor analogy but perhaps helpful.

6. Natural Selection

The words are out of the bag—evolution and selection. Are they necessary for genetics? The answer is a cautious yes and no. When Gregor Mendel studied what we now call genetics, he left open the possibility of hybrid stability in some species, but not in *Pisum*. He proposed a mechanism for the recombination that results in new forms, which ultimately gain stability through homozygosity. On the other hand, it seems pretty clear that Darwin would have accepted much of what Mendel wrote, had he been familiar with Mendel's work—contrary to what those who like to paint them in different camps would like to believe. Chapter 8 ("Hybridism") of *Origin of Species* is very closely related to Mendel's work. In fact, some passages in this Chapter were marked by Mendel in his personal copy; they are remarkably similar to Mendel's own work. So evolution and genetics are closely related.

In general, one can probably study genetics without accepting evolution, but it is harder to study evolution without knowing some genetics. I personally think that the study of genetics has it right and that the study of evolution has it right as well. I am not saying they both have gotten all their final answers, for nothing is final in science—neither am I saying that you cannot study genetics if you do not accept evolution. But let us assume for now that you do accept evolution, as I do.

Natural selection works on phenotypes, but it has no effect if there is no genotype behind it, for it is the genes and its alleles that are passed on to the next generation. This is an important condition for natural selection to work. When farmers dock the tails of their sheep, they are not docking genes, so natural selection cannot select this feature for future generations. They have to keep docking! When dentists pull wisdom teeth, they are not pulling genes, so people from the next generation will struggle again with their wisdom teeth. For natural selection to have effect, the only traits that qualify for selection must be hereditary as well as genetic at the same time.

Natural selection promotes "good" genetic designs more so than "bad" genetic designs, which makes them increase their frequency in future generations. The fact, for instance, that the caterpillars of a white cabbage butterfly are green rather than white—which is caused by certain genes—makes these slow organisms feeding on green cabbage less conspicuous to predators and thus more successful in survival and reproduction. In cases like these, we are not so much interested in the causes of camouflage (such as genes) as we are in its effects on survival.

Hence the reasoning goes basically as follows: Their green color causes camouflage; camouflage is a successful genetic design; therefore natural selection causes this functional causality to spread. In other words, the green color of caterpillars has a selective advantage over other colors, and therefore increases its frequency through better chances of reproduction. The allele for a green color confers a relatively better fitness in this environment and gives those caterpillars a selective advantage, which is taken as a positive selection pressure. Biologists usually speak in terms of *functionality*: A functional design is successful in survival and reproduction.

In other words, natural selection favors biological designs that are functional by solving a problem posed by the environment. Natural selection favors causes that have "successful" effects, so we call those effects functional. Darwin's theory of natural selection is based on the fact that organisms are genetically different with regards to their chances of survival and reproduction—although he did not know the genetics behind it. The more an organism is adapted to its environment—making for a better design fit to solve a problem posed by the environment—the more likely this organism is to contribute to the genetic constitution of the next generations. To put it in a nutshell, success breeds success.

This seems to be a law of nature in Darwin's view. Somehow, natural selection promotes the better designs by "weighing" the benefits against the costs in the same way as an engineer, economist, or architect would. Of course, organisms do not really calculate costs and benefits, but natural selection makes them act *as if* they do make strategic decisions. The wing of a bird, for instance, follows the same laws of aerodynamics as the wing of a plane. Whatever follows the correct laws will likely succeed.

The aim of Darwin's theory of natural selection is to explain any trait in a population in terms of *adaptation*. Evolution is supposed to produce better adapted, more functional phenotypes. If a genetic trait X has a function Y, then Y is called "adaptive"—which means that it promotes survival and reproduction. And therefore X will become an adaptation, for any trait having a function increases the likelihood that organisms with that trait will survive longer and/or reproduce more in comparison with organisms with less adapted traits. Consequently, a trait is not an adaptation in itself; it is rather an adaptation in comparison with other traits in a particular environment and with respect to particular criteria.

Functionality can be measured in terms of *fitness*—being the expected genetic contribution to future generations. The ultimate function of a trait is to enable organisms to adapt to their environment, to increase the chances of individual survival and reproduction—briefly, to enhance fitness. Given the fact that natural selection is a process of selecting better and better solutions, if available, for the problems a population encounters in its environment, one can construct an "optimal design" which can be set alongside the actual solutions that a certain population has developed.

This is called an "optimality model." Such a model takes the natural world as if it were designed by an engineer or economist who is concerned to get the maximum output for the minimum input. An optimal foraging model, for example, tries to show that particular foraging patterns maximize or optimize the net calorie intake of organisms; it assumes that organisms will forage first for food items that give the greater harvest per unit of time.

As already discussed, organisms do not calculate costs and benefits, rather they act *as if* they make strategic decisions. Eye patterns on butterfly wings, for instance, have the effect of warning enemies; that is a function of eye patterns, not a purpose of butterflies. Thus, optimality modeling is a way of finding out whether populations solve a given environmental problem in the way the optimality model predicts. When there is a discrepancy between optimal and actual solutions, there are plenty of reasons why the optimal solution may have been compromised—for instance, genetic and developmental constraints, past history, opposing selection pressures (sexual selection versus predator selection), and many other factors.

Some have attacked the idea of natural selection by arguing that it amounts to a tautology—that is, a statement which is true in every possible interpretation. Their argument goes like this: "Who survive? The fittest! Who are the fittest? Those who survive!" I do not think we need to take this attack seriously, for it is based on some terminological confusion. Biological fitness has actually a double meaning: It refers to the role of an organism as the subject, or *cause*, of reproduction, but also to the role of an organism as the object, or *effect*, of reproduction.

This ambiguity also affects the fitness concept: Darwin's concept of fitness—let us call it D-fitness—is potential reproductive success (a cause), but there is also the concept of fitness in the sense of actual reproductive success (an effect), which is the way the statistician Robert Fisher and some population geneticists use it—let us coin it F-fitness. Apparently, the slogan "survival of the fittest" is not a tautology if we take fitness as D-fitness, or reproductive capacity. The principle of natural selection (or survival of the fittest) asserts that those organisms that are potentially successful in reproduction (D-fitness) are more likely to be also actually successful in reproduction (F-fitness). Thus having a good design does matter in evolution.

As to the question whether natural selection is the only factor operational in evolution, the answer is plainly negative. It may be a main factor, but this is not to say that there are no other additional factors involved, such as mutation, migration, isolation, sampling, catastrophes, random-walk (genetic drift), and the like. Nevertheless, what matters the most here is that Darwin's theory opened the gate to a very successful research program. Because of this program, Bernard Kettlewell began to study how the composition of the peppered moth population experienced a change when industrial areas in England became more polluted. Then H.W. Bates started the study of mimicry— the phenomenon that an unprotected species (say, of flies) takes advantage of its partial resemblance to a protected species (say, of wasps)—and then he discovered that when the protected species varied geographically, its mimicking satellites had undergone exactly the same changes as their unpalatable models. Then there was Georges Teissier who designed simple population cages to experiment with different kinds of selection pressure on fruit flies. Later on, we see population geneticists using gene-pool models that they subjected to specific selection pressures (see Chapter 5). As a matter of fact, Darwin's research program was so powerful that many biologists to this day keep successfully working within its framework.

In short, what Darwin did—and what made him revolutionary—is that he approached all aspects of life as natural phenomena, which are to be explained by natural causes embodied in objectively testable theories. As Charles Darwin once wrote in one of his letters, "astronomers do not state that God directs the course of each comet and planet." He was right on target: Comets and planets just follow laws. Why would the same not hold for the process of evolution? Evolution just follows laws. (As to where those laws come from, that is another question.) Thus, modern biology was born. I consider this a great part of Darwin's legacy, but it is probably not the end of the story. Only time will tell.

7. What Selects What?

In 1859, when Charles Darwin published his best-known book—*On the Origin of Species by Means of Natural Selection, or the Preservation of Favoured Races in the Struggle for Life*—not only did he argue that all organisms descended from previous generations by gradual modifications, but he also suggested a causal mechanism for this process by embracing the concept of natural selection (see Chapter 6). Curiously enough, in his book, he did not use the word "evolution," but always referred to it as "descent with modification." However, Darwin did use the verb "evolve." In fact, this verb is the last word in the book, in Darwin's famous last sentence: "endless forms most beautiful and most wonderful have been, and are being, evolved."

Soon, however, he was accused of replacing God as a selecting agent with nature as a selecting agent—something like "the goddess Nature." Some say he just substituted a divine watchmaker with a "blind watchmaker." Darwin must have worried about this issue himself, for he always felt uneasy about his term "natural selection." As a matter of fact, this term leads us almost automatically to the obvious question "Selection by whom or by what?"

Darwin certainly tried to avoid this implication by saying he had as much right to use metaphorical language as physicists do. In his own words, "who objects to an author speaking of the attraction of gravity as ruling the movements of the planets? Everyone knows what is meant and is implied by such metaphorical expressions." Yet, his colleague Alfred Wallace convinced Darwin to replace the term "natural selection" with Spencer's idea of "survival of the fittest" in the 5th edition of his book. Darwin also liked the term "natural preservation."

No matter which terminology you choose or prefer, the question of "what selects what?" remains pressing. Darwin had the impression that he had replaced what William Paley had called a "Watchmaker," a purposeful Designer, with a purely causal, purposeless mechanism. In a sense it is true that Darwin's theory of natural selection attempts to explain nature's beautiful design in terms of physical causes and natural laws—without invoking Designer interventions—for he was always very well aware of the fact that science should search for a lawful evolutionary mechanism similar to those mechanisms many astronomers and physicists had come up with before him in their respective fields. He explicitly followed the philosopher John Stuart Mill's rule (1872) that says "it is a law that every event depends on some law." What a great insight Darwin had: Comets and planets do follow natural laws— and something similar must be the case in evolution.

In another sense, though, Darwin did not succeed in his endeavor, as he ignored the following question: Why do certain biological designs "work," and why are they "successful" and "effective" in reaching their "goal"? What is it that makes them "fit" to different degrees? Fit for what? Or put differently: What carries them through the filter of natural selection? I admit, these are not physical but meta-physical questions, located in a "meta" realm beyond, behind, beneath, outside, or underlying the regular physical realm—but that is not to say those questions do not exist. Yet, as a scientist, Darwin decided to leave such questions untouched. Is that possible?

Even if we acknowledge that a heart works because it is designed like a pump, and that an eye works because it is designed like a camera, we still have to face the following philosophical, meta-physical question: How come that a pump in itself "works" at all, and that a camera "works" at all? In other words, natural selection may explain that a fine working design has a better chance of being reproduced, but ultimately it cannot explain why such a design is working at all, let alone working so well. That is where *teleology* keeps creeping in, given the simple fact that the term "design" is a teleological concept in itself. Teleology embodies the idea that nature has "goals" and does things "in order to..." or "for the sake of..." Each time we are speaking of being "fit," "successful," or "goal-oriented," we are actually talking teleology—like it or not. Darwinism could not survive without teleology. Let me explain.

The issue is this: A heart and a pump do succeed because they work in accordance with the laws of nature; they follow the laws and constraints laid down in what I like to call the *cosmic design* of this universe. The universe has an overall set of restraints harnessing individual designs and making them "fit" or "successful" to a certain degree. Without a meta-physical design in the "background," biological as well as technological designs could not work at all. A heart could not pump blood if it did not follow hydrodynamic laws; a bird's wing would not let the bird fly if it did not follow aerodynamic laws. To use a technological analogy, the most perplexing thing about a watch is not so much that someone invented such a design but that the universe at all allows for any kind of design to work the way it works, thus making certain designs more goal-directed and more successful than others— which is a matter of teleology again.

George Bernard Shaw once said that Charles Darwin had thrown Paley's "watch" into the ocean. Well, Shaw was wrong. What Darwin did throw away was Paley's "watchmaker," but certainly not his legendary watch; if he did throw something away, it was Paley's design-Designer, but not the design concept itself. Make no mistake, *design* is an artifact analogy that is as basic to Darwinism as it is to Paley's natural theology. Since the heart is designed like a pump, it is a successful design "for" circulating blood. Apparently, Darwin did not discard design or what comes with it, teleology; after Darwin, the heart still existed "for" circulation; the cause of its existence may have been different, but its teleology was not.

So where do purpose, goal, and design come from then? Somehow they must have been built into nature—as some kind of all-pervasive, goal-directed architecture, as some kind of cosmic design. Apparently, there is "something" in successful biological designs that carries them through the filter of natural selection. In other words, natural selection on its own cannot do the "job" unless it works within a framework of cosmic design. Without this cosmic design working "in the background," there could be no selection. Natural selection can only select those specific *biological* designs that follow the rules of the *cosmic* design. Needless to say that designers, engineers, and architects must do the same thing.

It is the "rules of the cosmic design" that restrict the range of possible end results. Birds must follow the same aerodynamic laws as planes— otherwise they surely fall from the sky. To put it in a nutshell, organisms do not have successful designs because they have survived; on the contrary, they have survived because their biological design squares well with the cosmic design. They must have "something" in their biological design that carried them through the filter of natural selection.

Darwin may have thought he could reduce teleology to mere causality, but his causality mechanism of natural selection can only work on condition that there is teleology in nature. Darwin may have believed that he took teleology out of science by explaining it in terms of a "physical mechanism"—and in a way he did—but at the same time he left it in as an untouched philosophical presupposition. Some have put it this way, organisms are not teleological because they have survived; on the contrary, they have survived (in part) because they are teleological. In other words, the causality of natural selection does not explain teleology, but assumes it. Natural selection operates upon what exists but offers no explanation for all that exists. Natural selection does not produce anything new but only filters products that exist already.

It is the cosmic design that explains teleology by determining which designs are possible, and then it "selects" those designs that fit best.

The fit survive because of the order and harmony between them and their environment—which is a matter of cosmic design. In other words, Darwin had it backwards or upside down. He thought he could explain nature's functionality with the process of natural selection, but instead natural selection selects what is functional within the setting of the cosmic design, so it can only be explained by assuming a cosmic design.

Let me summarize all of this one more time. Due to the cosmic design, there is enormous potentiality in nature. It determines what is possible—and what is not possible—in this universe. All biological designs have to go through the "filter" of natural selection. Only what is in accordance with the *cosmic* design can be a successful *biological* design—successful in reproduction and survival. Natural selection can only select those specific biological designs that follow the rules of the cosmic design (again, designers, engineers, and architects must do the same thing).

Needless to say that genetics too must deal with this cosmic design. If hemoglobin does not have the "right shape," it causes red blood cells to "sickle"—which makes it a rather unsuccessful biological design, as it violates the rules and laws of the cosmic design.

8. Can Things Go Wrong?

If it is true that certain features are more "fit" or "successful" than others—as we suggested in the previous chapter—one could pose the following question: Can things go wrong in terms of genetics? The answer to this question is partially yes and partially no.

Let me start with the yes-answer first. Certain biological designs are less "fit" or "successful" than others in terms of natural selection. Consider the case of the formation of sperm cells and egg cells, when all pairs are split in half so the new cells have only one set of 23 chromosomes. If the chromosome pairs *fail* to separate properly during cell division, the egg cell or sperm cell may end up with an extra copy of one of the chromosomes, giving the resulting embryo three copies instead of one (*trisomy*).

What causes chromosomes to split improperly is not well understood. In general, this may happen more with increasing age of the mother. It can happen to any of the 23 chromosomes, but in most cases this leads to an early miscarriage. Estimates are that even more than 50% of abortions occurring spontaneously in the first trimester of pregnancy are caused by chromosomal aberrations. However, some triple-state chromosomes, so-called trisomies, do allow the organism to survive after birth. It may represent a handicap, but many of these people are able to live a happy and healthy life despite the challenge.

Some chromosomes—numbered or labeled as 21, 18, 13, X and Y—are seen in live born children as full trisomies, while trisomies of the chromosomes 15, 16, and 22 are often seen in miscarriages. The other chromosomes—1 to 12, 14, 17, 19, and 20—are almost never seen as full trisomies. Let us discuss in more detail the most common types of trisomy that survive to birth.

Trisomy 21, or Down syndrome, is the most common chromosome

abnormality in humans. It is typically associated with some delay in cognitive ability and physical growth, and a particular set of facial characteristics. In spite of the fact that the average IQ of young adults with Down syndrome is lower than that of children without the condition, many children with Down syndrome do graduate from high school and are able to do paid work, and some participate in post-secondary education as well. Education and proper care have been shown to improve quality of life significantly. And they are known to have a magnetic personality.

After Down syndrome, the second most common form of trisomy that carries to term trisomy 18, or Edwards syndrome. It occurs in around 1 in 6,000 live births and around 80% of those affected are female. The majority of them die before birth. This syndrome has a very low rate of survival, resulting from heart abnormalities, kidney malformations, and other internal organ disorders. Yet, about 8% of these infants survive longer than 1 year. Although women in their 20s and early 30s may conceive babies with this syndrome, the risk of conceiving a child with this syndrome increases with a woman's age.

Then there is trisomy 13, or Patau syndrome, which causes these infants to have difficulty surviving the first few days or weeks due to severe neurological problems or complex heart defects. Treatment focuses on the particular physical problems with which each child is born. Surgery may be necessary to repair heart defects or cleft lip and cleft palate. Physical, occupational, and speech therapy will help individuals with this syndrome reach their full developmental potential.

Trisomy XXY, or Klinefelter syndrome, is a condition in which an XY-person has an extra X-chromosome. Since these individuals do have a Y-chromosome, they will develop the male sexuality and are therefore usually referred to as "XXY males." This form of trisomy occurs in roughly 1:500 to 1:1000 live male births, but many of these people may not show any symptoms. The physical traits of the syndrome, including reduced strength and reduced fertility, may become more apparent after the onset of puberty, if at all.

Trisomy XXX, or Triple-X syndrome, causes a female to have an extra X chromosome in each of her cells. Because the vast majority of Triple-X females are never diagnosed, it may be very difficult to make generalizations about the effects of this syndrome. The reason why extra X-chromosomes have hardly any effect is that in female cells only one X-chromosome is active at any time, whereas additional X-chromosomes are kept inactivated and remain condensed into so-called Barr-bodies. No wonder that balanced information provided to prospective parents, before birth, has been shown to reduce the decision to terminate the pregnancy.

The mechanism of inactivating additional X-chromosomes is not fully understood. It is somehow regulated by a gene on the X-chromosome called XIST (X-inactive specific transcript). When this gene comes to expression, it produces a large non-coding RNA that "silences" the X-chromosome from which it is transcribed by "coating" this chromosome with RNA molecules so it becomes inactivated. It has been found that artificially placing and expressing the XIST gene on another chromosome leads to inactivation of that chromosome as well. This has given researchers the hope that someday they might be able to also "silence" other redundant chromosomes in trisomy cases such as Down syndrome.

And then there is monosomy XO, or Turner syndrome. It is like the opposite of trisomy—instead of three chromosomes, there is only one, with the second one missing (O). In 75% of the cases, the X-chromosome came from the mother, so the father did not contribute an X-chromosome. Girls with Turner syndrome typically have dysfunctional ovaries, which results in sterility and having no menstrual cycle. Often there are also health concerns, such as congenital heart disease, reduced hormone secretion by the thyroid, diabetes, vision problems, hearing concerns, and many autoimmune diseases. Hormone treatments are possible, though. It has been found that providing very low doses of estrogen to girls with Turner syndrome as well as growth hormone, years before the onset of puberty, increases their height and offers a wealth of other benefits.

A complication in the diagnosis of trisomy cases is that the third chromosome may not be a complete chromosome copy but only a partial one. In trisomy 9p, for instance, the third copy only consists of a portion of the short arm of chromosome 9. On the one hand, virtually all individuals with Trisomy 9p are affected by some mental retardation and distinctive malformations of the skull and facial region. On the other hand, trisomy 9p is one of the most frequent anomalies that has a long survival rate.

It should be noted, though, that all the trisomies mentioned here may also develop *after* conception during the production of new cells, so it may only show up in a particular cell line. In such cases, only some of the body's cells have an extra chromosome copy, resulting in a mixed population of cells with a differing number of chromosomes. Such cases are sometimes called *mosaic* syndromes. Very often the mosaic version of trisomy exhibits a very mild range of physical abnormalities and developmental delay compared to full-trisomy cases.

The cases we mentioned so far lead to an irregular number of chromosomes. But even if the number of chromosomes is normal, they can obviously carry alleles that have a negative impact on survival. The list could be quite extensive. Sometimes the effect of such alleles is minor because they are recessive and can be compensated for by its paired allele. If not, the negative effects may need extra attention through the intervention of surgery, medication, and/or special nutrition.

Nowadays, a more "invasive" treatment may be *gene therapy*. Gene therapy is the use of DNA as a pharmaceutical agent to correct the wrong allele and/or its protein. Conditions that arise from mutations in a single gene are the best candidates for gene therapy. The most common form of gene therapy involves the insertion of a functional gene in the host genome at a specific location in the body. This is accomplished by isolating and copying the gene of interest, adding all the genetic elements for correct expression, and then inserting this assembly into the host organism. In order to get the therapeutic DNA inside the cell, the most common and successful method is the use of recombinant viruses by removing the viral DNA and using the virus as a vehicle to deliver the therapeutic DNA.

So the DNA that carries the code for a therapeutic protein is packaged within this "vector" in order to get the DNA inside cells within the body. Once inside, the DNA becomes expressed by the cell machinery, resulting in the production of therapeutic protein, which in turn treats the patient's disease. Unfortunately, there is always the fear that the viral vector or plasmid, once inside the patient, may recover its ability to cause disease. Besides, if the DNA is integrated in the wrong place in the genome, for example in a tumor suppressor gene, it could induce a tumor.

Nevertheless, there have been some success stories lately. Recent clinical successes include treatment of patients with chronic lymphocytic leukemia (CLL), acute lymphocytic leukemia (ALL), multiple myeloma, hemophilia, and Parkinson disease. In all these cases, the therapeutic genes were transferred into the somatic cells (non sex-cells) of a patient. Therefore, any modifications and effects will be restricted to the individual patient only, and will not be inherited by the patient's offspring or later generations (in fact, benefits are usually short-term, so patients will have to undergo multiple rounds of gene therapy). This is called *somatic cell* gene therapy.

In addition, there is also *germ line* gene therapy, which modifies sperm cells or eggs cells by the introduction of functional genes, which are integrated into their genomes. This would allow the therapy to be heritable and passed on to later generations. No wonder, many countries (not the USA, though) prohibit this method for human beings, at least for the present time, because of technical and ethical reasons, most importantly insufficient knowledge about possible risks to future generations. Even the *somatic* version of gene therapy has still many uncertainties and risks, including the risk that the new gene crosses the barrier between body and germ line by spreading to the testes or ovaries.

So as to the question whether things can go wrong for the next generation, the answer is yes. But let us not forget that there is no "perfect" genome. Everybody has genetic flaws. It just depends on how they become manifest in someone's life. Hemoglobin in its regular form, for instance, has a "perfect" functional design to transport oxygen in the body, but it becomes "dysfunctional" as soon as the body gets exposed to carbon monoxide. If you are looking for something wrong, you will certainly find it. In that sense, the answer is yes—things can go wrong.

However, seen in a different context, the answer could also be a definite *no*. Things can be looked at from different perspectives. Soon after the debut of classic movie "How the West Was *Won*," another one was made from a different perspective: "How the West Was *Lost*." They each had a different part of the story to tell. Think of something like walking around a statue; what we see in the back is different from the front, not so because we *wish* the back to be different but because the back *is* actually different from the front. In a similar way, there are many different perspectives and outlooks on the world surrounding us.

When we say "things went wrong," we could mean two entirely different things: "wrong" can be taken in a descriptive, technical sense, but also by way of evaluation and moral assessment. What may look "wrong" from a biological perspective, may not be "wrong" in any other sense. When scientists tell us that mutations can be good or bad, they use, or should use, the descriptive version—some mutations enhance functionality, others lessen functionality. All that scientists can say is that mutations are a matter of randomness or coincidence and confer more or less functionality under certain circumstances, but never could they speak of "fate" or "luck," let alone "good or bad luck."

The notion of "luck" does not belong to a scientist's vocabulary—it is a worldview notion, associated with concepts such as doom, fate, and misfortune (versus destiny, blessing, and divine intervention). When we speak of "bad luck," we consider something as a failure, as something that we dislike, condemn, or even reject—in short, as something that *should not* be. However, judgments about what should be and should not be lie outside the reach of science and go far beyond a scientist's competence—they are actually *moral* evaluations and judgments, often with a religious or semi-religious undertone (see Chapter 15).

In other words, a process may have gone wrong in a genetic or biological way, but that does not mean it was a failure in a wider context. Changes in genetic material may be random, but that does not make them a matter of "good or bad luck." Since genes always come in duplex, one of the two goes to the next generation—but as to which one, that is supposedly also a matter of chance. When George Bernard Shaw was approached by a seductive young actress who cooed him in his ear, "Wouldn't it be wonderful if we got married and had a child with my beauty and your brains," Shaw replied: "My dear, that would be wonderful indeed, but what if our child had my beauty and your brains?" Yes, genetics can go either way, because it works with randomness and probabilities.

Randomness is a scientific term that tells us is that causes can be in-dependent of each other. It has a narrow technical meaning having to do with the statistical correlations among things and allowing us to calculate probabilities. Events do not happen "by chance" because they are uncaused or we cannot trace their causes, but because there are so many causal chains that intersect with each other. Whereas randomness and chance describe the relationship *between* causes, luck is a worldview notion that refers to some kind of agency *beyond* the realm of regular causes. From this follows one very important ca-veat: Never confuse randomness with fate, luck, or doom, as these latter terms are not scientific concepts but world-view notions.

Let me use a comparison. When things go "wrong" in someone's life, career, or marriage, that does not mean this person is a "lost case" who does not deserve to live in our society. Genes do not determine whether we are a failure or not; they are like a hand of cards we are dealt, but we can play them differently. We can lose with a hand of "good" cards, but even with "bad" cards, we may be able to win. Our genes do not determine our destiny. When problems loom, so do op-portunities.

Nevertheless, some scientists confuse "normal" taken as a *de*scriptive term with "normal" as a *pre*scriptive notion. Whereas the word "nor-mal" can *de*scribe what the average is, the same word can also *pre*-scribe what is considered normal versus what is deemed "abnormal." Yet, some scientists ignore this difference when they go beyond their description of the way things *are*, and then declare what way things *should*, and should not, be. My response would be: Why would we consider some mutations more equal than others? Who has the right to claim that there are some human lives unworthy of living? Who has the authority to make us a judge of life and death?

Unfortunately, the biological notion of "survival of the fittest" has in-doctrinated many biologists and physicians to assess everything in terms of "winners" and "losers"—only the "fittest" are supposed to survive. They misguidedly jump from a descriptive notion to a pre-scriptive notion. But are people with Down syndrome or any other chromosomal "disorder" really "misfits" or "losers"? They only would be if we place the final end of human beings solely in their biological worth. But is there not much more to life than biological worth? Is there not much more to life than genes? Just ask people with mental or physical disabilities whether they feel worthless. If they do say yes, it is most likely other people who made them feel that way.

Nevertheless, many scientists would tell you that chromosomal aber-rations and other kinds of changes in genetic material are "random," and are therefore a matter of pure luck—in their view, a form of "bad" luck that we should eradicate. For now, I will only say that some scien-tists talk about "random" as if it were a deity with a capital R—but such a concept is more like a world-view notion, certainly not a scien-tific concept. What they have in mind is the goddess of fate or doom. But there is no space for deities in science—or at least there should not be. Those who still maintain that concepts such as fate and doom belong to science are no longer teaching science but preaching sci-ence, disguising themselves as doom-thinkers, as prophets of cyni-cism. Those who confuse "random" with words such as "ill-fated" or "unlucky" may be scientists but they are no longer speaking scientifi-cally.

Who is going to decide which newly discovered mutation goes on the list of "unwanted" alleles? Most decisions concerning congenital mu-tations that condemn human beings—born or unborn—to destruction are arbitrarily made behind the closed doors of hospitals and pharma-ceutical companies. Who gives anyone the right to make such deci-sions? As we found out, everybody has genetic flaws and is in essence a walking genetic "junkyard." If you are looking for something wrong, you will certainly find it. Even some people who herald diversity in our society end up going for the uniformity of the "fittest."

Weeding out what is "genetically flawed" is certainly not a matter of genetics. Genetics can describe but not prescribe. The belief that some humans command a favored right to life—simply because of their physical or genetic "fitness"—is not based on science, but ideology. It is actually the ideology of eugenics. Eugenics basically asserts that we should breed humans like we breed animals—hence we should be able to kill them like we kill animals.

After it started in the late 1800s, eugenics soon developed into a brutal movement which inflicted massive human rights violations on millions of people. The "interventions" advocated and practiced by eugenicists involved a wide range of "degenerates" or "unfits"—the poor, the blind, the mentally ill, entire "racial" groups such as Jews, Blacks, and Roma ("Gypsies"). All of these "misfits" were deemed to be "unfit" to live according to their despotic dogma called "survival of the fittest." This, in turn, led to practices such as segregation, sterilization, genocide, euthanasia, pre-emptive abortions, designer babies, and in the extreme case of Nazi Germany, mass extermination. G.B. Shaw once predicted that "part of eugenic politics would finally land us in an extensive use of the lethal chamber."

No matter how you look at it, these "eugenicists" have charged themselves with the grave duty to decide who is to live and who is to die—a quasi-moral stand based on biological grounds. Instead, I think we need to keep stressing that no one has the right to claim that there are some human lives unworthy of living. Ironically, those who claim this right often fail to cope with many other equally important issues in life, whereas those who "should not qualify for life" may end up learning to cope with their medical adversity in unexpected and surprising ways. What some call a curse may turn out to be a blessing. Each setback opens the prospect of a comeback. Think of what you would have missed if one of your dear ones with a genetic disease would not have been given a chance to live. If you have a disease, let the disease not have you. When a blind woman was asked how she could be so joyful with her blindness, she responded with the question "How can you see and *not* be joyful?"

III. The Double Helix

9. From Genes to DNA

The history of genetics has shown us how the genes of genetics gradually became more and more concrete entities (see Chapter 1). They evolved from the abstract hypothetical "elements" Mendel spoke of, to the more concrete "factors" Bateson referred to, then to what Johansson called "genes," and finally to the "beads" on Morgan's chromosomes. Yet, for a long time, they were still taken more as accounting or calculating units than as material entities. This changed drastically when James Watson and Francis Crick introduced their DNA model in 1953. Since then, our understanding of the material basis of the gene has gone through an accelerating growth process.

What is it that the genes are made of? The simplest answer is DNA. A molecule of DNA (*Deoxyribo-Nucleic-Acid*) is composed of building blocks called *nucleotides*, each of which is itself composed of a *five-carbon* sugar (ribose) bonded to a phosphate group and a nitrogenous base. For the hungry minds among us: There are four kinds of nucleotides in DNA, which differ from one another in their nitrogenous bases: adenine (A) and guanine (G) are double-ring structures, whereas cytosine (C) and thymine (T) are single-ring structures. In a DNA molecule the nucleotides are arranged in sequence, held together by covalent bonds between the sugar of one nucleotide and the phosphate group of the next nucleotide beyond it; the nitrogenous bases are arranged as side groups off the chains.

DNA molecules ordinarily exist as *double*-chain structures—comparable to a zipper—with the two chains held together by hydrogen bonds or bridges between their nitrogenous bases; such "bonding" can occur only between a single-ring and a double-ring structure—that is, between C and G (C-G bonds) or between T and A (T-A bonds). Finally the ladder-like double-chain molecule is coiled into what is called a helix, actually a double helix.

For a while, geneticists held on to what they literally called their "central dogma" which says: "DNA makes RNA makes protein"—that is, a one-directional causal flow, with one item of code, a gene, ultimately making one item of substance, a protein, by forming a sequence of amino acids, and with all these proteins together making a body. Since there are 4 different nucleotides (A, C, G, and T) and 20 different amino acids, the coding unit of DNA, a *codon*, must be 3 nucleotides long, which makes for 64 possible combinations (4^3) and thus leaves room for "synonyms" (e.g. the codons GCA, GCC, GCG, and GCU all specify the same amino acid, alanine).

DNA acts as a template for the synthesis of messenger RNA (mRNA). A DNA code such as CGG would be transcribed into the complementary, "anti-sense" mRNA code as GCC, because C and G can bond, as do A and T. This mRNA, in turn, determines the order of amino acids in a polypeptide (enzymes and other proteins). If you want to know what the more detailed sequence of steps is, keep reading the rest of the next three rather technical paragraphs—otherwise just skip them.

The two strands of a DNA molecule in the cell nucleus first un-couple so that one of the two strands can act as a template for synthesis of a complementary, "anti-sense" string of mRNA. Then mRNA leaves the nucleus and goes to the cytoplasm, where it complexes with ribosomes. There is also tRNA which carries one specific amino acid to mRNA at a time. tRNA couples briefly with mRNA on a ribosome, and then moves off to pick up a new amino acid. While the ribosome moves along on mRNA, it adds amino acids to the growing polypeptide chain, which finally results in an enzyme or any other protein.

The direction in which DNA is transcribed into RNA is very different from the direction in which DNA is synthesized into a complementary DNA strand. When nucleotides are joined together, the "first" one still has a free phosphate group attached to the 5th carbon (C5) of the sugar ribose, but it has lost its –OH group on the 3rd carbon (C3). The "second" nucleotide has its C5 phosphate now bound to C3 of the first nucleotide. These C5-C3 bonds are repeated many times as the nucleotides string together to form a long DNA strand. When the strand has been completed, both ends of the strand look very different. One end still has a free C5 phosphate group—this is called the "5' (upstream or head) end." The other end still has a –OH group on C3—called the "3' (downstream or tail) end."

The synthesis of DNA is done in the "5' to 3' direction"—which means the nucleotide chain is "read" from the 5' end to the 3' end, just like English is read from left to right. But RNA is written in the opposite direction, from 3' to 5', when being transcribed from the corresponding "anti-sense" DNA strand. So the "anti-sense" DNA strand *(5')* CGG *(3')* would be transcribed in the corresponding RNA strand as *(3')* GCC *(5')*.

However, the previous overview is still very simplified and has dramatically changed more recently. First, it turned out that some DNA sections, so-called *introns*, get initially transcribed into mRNA but are then removed from the end-product by so-called splicing. The largest human gene, dystrophin, a muscle protein implicated in muscular dystrophy, has many introns and is 2.4 million bases in length, which requires a lot of splicing. It is due to *alternative* splicing that a single gene may code for several different proteins. This would partially explain why the number of genes can be much lower than we had initially expected.

Second, genes can be overlapping; about 9% of human protein-coding genes overlap another such gene. Sometimes the overlaps are partial, but in other cases small protein-coding genes are fully embedded within much larger genes (e. g. the blood clotting factor VIII)—so they form genes-within-genes. In this way, a nucleotide sequence may make a contribution to the function of more than one gene product. In other cases, some DNA sequences can even do double duty, encoding one protein when read along one strand, and a second protein when read in the opposite direction along the other strand. All these cases may very well represent a hidden source of complexity to modulate gene expression.

Third, it turned out that protein-coding regions of genes can be interrupted by DNA segments that play more of a regulatory role. These regions are either called *enhancers* or *silencers*. They are short regions of DNA that can be bound with proteins to enhance or silence transcription levels of genes in a gene cluster. Enhancers produce *activator* proteins that activate the activity of a "regular" gene, whereas silencers produce *repressor* proteins that repress gene activity. In either case, they produce DNA-binding proteins that regulate gene expression by binding to a segment of DNA called the *operator* that normally initiates gene transcription and is located near the gene it transcribes. In other words, enhancers and silencers do not act on the promoter region itself, but are bound by activator or repressor proteins.

Fourth, some regulatory DNA segments are actually very short and produce only short strands of mRNA capable of blocking the mRNA of a "regular" gene from creating its own regular protein; these DNA segments are called micro-RNAs (miRNA). Predictions are that they control the translational activity of approximately 30% of all protein-coding genes in mammals.

Fifth, some genes have actually lost their functionality—they are called pseudo-genes. These genes resemble a regular DNA packet of a functional gene, but they have been affected by one or more glitches that change their script into "nonsense." They were once functional genes but have since lost their protein-coding ability—and, presumably, their biological function. When comparing humans and chimpanzees, for instance, we do find genes that are functional in one species but not in the other—so we call them pseudo-genes. A striking example is the gene for a jaw muscle protein (*MYH16*), which has become a pseudo-gene in humans, but is still functional in developing strong jaw muscles in other primates. Another example would be the DNA sequence for an enzyme that produces ascorbic acid, or vitamin C, in most animals. Many primates, including humans, have a defect in this DNA code, so they must acquire vitamin C through food—yet they hold on to this pseudo-gene.

And then there has been a sixth important development. Genes may be separated by long stretches of DNA that do not seem to be doing much at all—that is why they are often called "junk DNA" in the popular press. Currently, geneticists rather speak in terms of "non-coding," "neutral," or "silent" DNA. Some of this "non-coding" DNA is *repetitive* DNA, often replicated from regular, coding DNA, and this perhaps serving as a rich source from which potentially useful new genes may emerge during evolution (more on this in Chapter 11). Speaking of simple genetic inheritance, the story is getting more and more complicated.

Special mention should be made of the so-called *homeobox*, or *Hox*, genes. The morphological diversity of vertebrates, from humans to eagles, or from whales to snakes, evolved around a common set of developmental genes. Mammals, birds, and amphibians, for instance, share the same set of 39 Hox genes. It is these Hox genes that control the body plan of the embryo along the anterior-posterior (head-tail) axis. In vertebrates, the various Hox genes are situated very close to one another on a chromosome in groups or clusters. Interestingly, the order of the genes on the chromosome is the same as the expression of the genes in the developing embryo. Humans have Hox genes in four clusters: HOXA on chromosome 7, HOXB on chromosome 17, HOXC on chromosome 12, and HOXD on chromosome 2. These genes somehow regulate that all vertebrates have a similar build; they make for the unity underlying morphological diversity. Morphological diversity, on the other hand, rather results from changes in gene regulation.

The products of Hox genes are Hox proteins, which are transcription factors (proteins) that are capable of binding to specific enhancers or silencers where they either activate or repress genes; the same Hox protein can act as a repressor at one gene and an activator at another. After the embryonic segments have formed, the Hox proteins determine the type of segment structures (e.g. legs, antennae, and wings in fruit flies or the different vertebrate ribs in humans) that will form on a given segment.

It must be clear by now that the gene definitions we used before— either Morgan's definition of a gene as a unit of recombination or the classical definition stating that a gene is a unit of heredity that regulates a specific trait, feature, or characteristic of an organism—are no longer appropriate on a molecular basis. We are talking now about "physical" genes that consist of stretches of DNA with a physical beginning and end.[1]

The gene concept has apparently undergone quite a conceptual change. Starting as an element (1865, Mendel), it became a factor (1900, Bateson), then a unit of transmission (1909, Johannsen), as part of a chromosome (1915, Morgan), with a specific order of DNA bases (1953, Watson and Crick), contributing to the production of enzymes (1940, Beadle and Tatum), or of polypeptides in general (1957, Benzer), having a structural or regulatory function (1961, Jacob and Monod), and consisting of exons and introns (1977, Chambon). And the conceptual shift keeps going on and on. As a result, the gene has moved far away from what Mendel had envisioned.

---

[1]   For   the   current   human   genetic   nomenclature,   see   www.genenames.org.

Soon after the discovery of DNA, a physical gene was defined as a DNA sequence that encodes a protein. This is the so-called "one-gene-one-protein" definition. One of the problems with this definition is that we know nowadays of genes that do not code for proteins, since there are also genes for transfer RNA (tRNA), for ribosomal RNA (rRNA), and for a large group of small RNAs (miRNA)—but none of them really have coding regions. The transcript is the functional end-product, often after RNA processing.

So we need a better definition. Try this one: A physical gene is a DNA sequence that is transcribed to produce a functional product (protein or not). Let us discuss what this definition entails.

Genes are transcribed from an initiation site to a termination site. When the RNA transcript is finished, it undergoes an additional step called RNA processing. In that step, parts of the original transcript (from introns) are spliced out and discarded, so the final mRNA cannot be translated until RNA processing is completed. In our definition, introns would be part of the gene since they are transcribed.

But of course, there are always exceptions in genetics! That makes it even harder to come up with an all-inclusive definition of what a "physical gene" is. Here is a short list of those exceptions, in case you are interested.

1. There is also trans-splicing: There are examples of "genes" that are split into pieces. The transcript from one piece is joined to the transcript from another one to produce a functional RNA.

2. There is also overlapping: Since some "genes" overlap, a single stretch of DNA can be part of two or even three genes.

3. There is also RNA editing: In some cases the primary transcript is extensively edited before it becomes functional. It may even happen that nucleotides are inserted and deleted. In such cases, the original

"gene" requires the assistance of other "genes" to ensure a functional end-product.

4. Then there are operons: They are units of DNA containing a cluster of genes under the control of a single regulatory promoter. Promoters are regulatory sequences upstream of the transcription start site; they bind the enzyme RNA-polymerase that catalyzes the synthesis of the RNA chain. The genes are transcribed together into one mRNA strand, after which they are either translated together, or they undergo trans-splicing to create several strands of mRNA, each of which encodes a single gene product. In these cases it does not make sense to refer to the co-transcribed genes as a single "gene." Instead, we would identify the stretches of DNA that correspond to a single functional unit as the "gene."

No wonder it has become very hard to come up with a comprehensive definition of a "physical" gene. The boundaries of a gene have become rather blurred. Moreover, we still need more than a "physical" definition in all those cases where the molecular definition cannot help us— for example, when we can only define a gene by its phenotypic effects and not its physical structure (yet). In those cases, we should probably revert to Morgan's definition—a unit of recombination.

Yet, a "physical" definition has become more and more popular today. One of the reasons is that we know more and more about the sequence of genes in the human genome due to the *Human Genome Project*. It started as an international scientific effort to map all the genes on the 23 pairs of human chromosomes and, to sequence the 3.2 billion DNA base pairs that make up the chromosomes. Begun in 1990, the project was largely completed in 2000 when 85% of the human genome was decoded, and ended in 2003 with 99% decoded; detailed analyses of all the pairs were published by 2006.

HGP researchers have deciphered the human genome in three major ways: determining the order, or "sequence," of all the bases in our genome's DNA; making maps that show the locations of genes for major sections of all our chromosomes; and producing what are called linkage maps, complex versions of the type originated by Morgan Hunt in the early 1900s with his *Drosophila* research.

Upon publication of the majority of the genome in February 2001, Francis Collins, the director of the National Human Genome Research Institute (NHGRI), noted that the genome could be thought of in terms of a book with multiple uses. In his own words, "It's a history book - a narrative of the journey of our species through time. It's a shop manual, with an incredibly detailed blueprint for building every human cell. And it's a transformative textbook of medicine, with insights that will give health care providers immense new powers to treat, prevent and cure disease."

How is DNA sequencing done? Today, "dideoxy sequencing" is the method of choice to sequence very long strands of DNA. The procedure of DNA sequencing is a multi-step process. Before it can be sequenced, DNA needs to be purified from cells by breaking apart cell membranes and removing protective proteins from the DNA. Very large pieces of DNA, such as whole chromosomes or genomes, are cut into smaller pieces and stored in vectors (plasmids), which are larger pieces of DNA with the ability to be reproduced when placed in host cells such as bacteria. Each time a bacterium divides, the DNA vector placed inside is also copied. In this way, the target DNA can be multiplied exponentially and forms clones of copies.

The sequencing reaction itself consists of four steps. First, the double-stranded DNA is separated into single strands (denaturation), and a small starter piece of DNA, called a primer, binds to the template strand. In the extension step, a new DNA strand is made that is complementary to the template strand by using the bacterial enzyme DNA polymerase. DNA polymerase cannot start copying a template strand unless there is a small piece of DNA to start the extension process—this is why the primer was added in the previous step. The termination step is the key to the sequencing reaction. In addition to the four regular single nucleotides, the reaction mixture also contains small amounts of four dideoxy-nucleotides, which lack the 3'-hydroxyl group necessary for chain extension. Once in a while—by chance, based on its much lower concentration—a dideoxy-nucleotide will be incorporated into the growing DNA strand. Because it is missing the 3'-hydroxyl group, the dideoxy-nucleotide will prevent the DNA chain from being extended further. Since each of these four special nucleotides is labeled with a different fluorescent dye, a certain kind of laser can later detect them.

Since an actual sequencing reaction mixture contains thousands of DNA template strands, which are all being sequenced simultaneously, primed chains end up being very short, very long, and of every possible length in between. The newly synthesized DNA strands, each labeled at the end with one of four dyes, are now loaded onto a tiny capillary tube containing a viscous, gel-like material. As the strands emerge out of the bottom of the capillary they pass through a laser beam that excites the fluorescent dye attached to the dideoxy-nucleotide at the end of each strand. This causes the dye to fluoresce, or glow, at a specific wavelength, or color. This color is then detected by a photocell, which feeds the information to a powerful computer that does the rest of the work.

The computer analyzes the overlapping sequence data from many clones and regenerates the full-length sequence by piecing the short sequences together like a puzzle. It also displays the information received from the photocell as an electro-pherogram, by printing the letter of the appropriate base below each of the signal peaks. Because successive peaks correspond to DNA segments differing in length by *one* nucleotide, the sequence of peaks reveals the sequence of bases in the original DNA sample. Entire genomes can be sequenced in this manner.

Instead of sequencing the entire genome, a cheaper method is sequencing only the exome, the part of the genome formed by exons. Exons are short, functionally important sequences of DNA which represent the regions in genes that are translated into protein. So these are the sequences which remain within the final RNA after introns are removed by RNA splicing. In the human genome there are about 180,000 exons; they constitute about 1% of the human genome. Though containing a very small fraction of the genome, mutations in the exome are thought to harbor 85% of disease-causing mutations. So exome sequencing is a cheaper but still rather effective alternative to whole genome sequencing

Not surprisingly, the Human Genome Project has caused quite a stir. It made many people think—unjustified, of course—that DNA holds the script for a person's entire life, which made them want to know their "personal genomics." The science journal *Nature* listed "Personal Genomics Goes Mainstream" as a top news story of 2008. The new genomics has now become our latest crystal ball.

10. Changes in DNA

By now, you must be wondering how differences between the alleles of a gene could have ever come along. The general answer is—mutations.

Mutations are random, permanent changes in the DNA sequence of a gene, thus producing a new allele. Mutations are the main source of genetic variability. Let me explain first what the words "random" and "permanent" refer to.

By "random" I mean several things. To begin with, mutations are considered random in the sense that they are—as far as we know—"unpredictable" as to when and where they strike. What we can predict is unpredictability. We may know more nowadays about what causes them—factors such as radiation, chemicals, free radicals, etc.—but we still cannot predict at what location in the DNA those factors will hit and what changes they might generate there. Second, mutations are random in the sense of "arbitrary," because mutations do not select their target but hit in-discriminatorily—"good and bad" spots alike. Third, mutations are also random in the sense of "opportunistic," because they occur without any connection to immediate or future needs of the organism. There is no physical mechanism that detects which mutations would be beneficial and then causes those mutations to occur.

But that is also where randomness ends. It certainly does not mean that mutations are random in the sense of being boundless or unrestricted. There are certainly limits to their effects—limits such as energy barriers and surrounding factors. As a consequence, some mutations may be more likely than others—yet they are random in the sense of what we discussed in the previous paragraph.

Randomness is just a statistical concept that allows us to calculate probabilities. It is like an empty shell—it does not even have a memory of past random events. When we toss a coin, we know it is going to be either heads or tails—but never the edge of the coin, for instance. If we had knowledge enough of the way the coin was flipped, we could probably predict the outcome exactly. But what makes the toss random is the fact that the outcome is not related to any particular player's needs. And it is also random in the sense that the next toss is not related to previous tosses—they are independent of each other. Even if we had a long series of heads or tails, the next toss is 50% head or 50% tails. That is why we can speak of statistical probabilities.

Let me add one more caveat: Never confuse randomness with fate or doom, as these latter are not scientific concepts but world-view notions. Those who confuse "random" with words such as "ill-fated" or "unlucky" may be scientists but they are not speaking scientifically. Randomness in genetics is a very technical concept, mainly borrowed from statistics and probability calculus and used when events are not connected or correlated with each other. In that sense, mutations are "statistically random"—they just happen one way or the other, but there is no correlation with other events such as environmental changes or immediate and future needs, so the former do not affect the latter, or reversed.

DNA can undergo damage as well as mutation, but these two are quite different. Damages are physical abnormalities in the DNA, such as single- and double-strand breaks in a chromosome, so the gene that the affected DNA encodes cannot be transcribed. DNA damages can be recognized by enzymes, and, thus, they can be correctly repaired if redundant information, such as the undamaged sequence in the complementary DNA strand or in a homologous chromosome, is available for copying. So organisms do have so-called mismatch-repair genes, which help recognize errors when DNA is copied before cell division takes place. Repair genes may kick in, fix the mismatch, and correct the error by using the unmodified complementary strand of the DNA as a template to recover the original information.

In contrast to DNA damage, a mutation is a change in the base sequence of the DNA. A mutation cannot be recognized by enzymes once the base change is present in both DNA strands, and thus, a mutation cannot be repaired. Mutations are replicated when the cell replicates. DNA damage occurs more often than you might expect; it is estimated to occur in as many as 1 million individual molecular lesions per cell per day. If a DNA copy does not "match" perfectly, errors in DNA can be transmitted to new cells and they become (permanent) mutations.

Genetic changes are caused by radiation, viruses, transposons, and mutagenic chemicals, as well as errors that occur during cell division or DNA replication. They can occur at the chromosome level or at the gene level; in either case, we may speak of a mutation, although most geneticists restrict mutations to DNA changes at the gene level.

At the *chromosome* level, entire chromosomes or parts thereof may get lost or amplified. A well-known example of this is the Philadelphia chromosome—a translocation in which parts of two chromosomes, 9 and 22, swap places after pieces of these two chromosomes break off and join together, forming an abnormal chromosome called the Philadelphia chromosome. This is associated with chronic myelogenous leukemia and results in production of an oncogenic enzyme called tyrosine kinase.

Something similar may also happen to entire chromosomes; they can get lost or amplified when the separation of chromosome pairs goes wrong during the formation of egg cells or sperm cells. A missing chromosome is usually detrimental, except for the X-chromosome (people with Turner syndrome (*XO*) have only one X-chromosome). Extra chromosomes are more frequent. Examples are people with Klinefelter syndrome, who have three X-chromosomes (*XXX*), or people with Down syndrome, who have chromosome 21 in threefold (more on this in Chapter 11).

More common, however, are mutations at the *gene* level, by changing the nucleotide sequence of nuclear DNA. They usually affect a single nucleotide, called SNPs (single-nucleotide polymorphism, pronounced "snips"). However, because codons have synonyms, such mutations may have no effect at all. If they do change what the codon codes for, however, the result may vary from minor to major. When they occur in the promoter region of a gene, they may affect whether the gene comes to expression or not. When they occur in the gene's coding sequence, they may alter the functionality or stability of its protein product. Disruption of a single gene may also happen when DNA material from a virus or retrovirus gets inserted into the gene, which can result in the expression of viral oncogenes in the affected cell and its descendants.

A particular SNP made headlines in the media recently. Sam Berns, a fan of the New England Patriots died of the disease, on January 10, 2014, at the age of 17 (had he lived another day, he would have served as the team's honorary captain in their playoff game versus the Indianapolis Colts). Progeria (or Hutchinson-Gilford progeria syndrome) is an extremely rare genetic disease wherein symptoms resembling aspects of aging are manifested at a very early age. It is a genetic condition that occurs as a new mutation, and is almost never passed on from affected parent to child, as affected children rarely live long enough to have children themselves. In 2003, the cause of progeria was discovered to be a point mutation in position 1824 of the LMNA gene, in which cytosine is replaced with thymine. This mutation causes transcription of the LMNA gene to stop too early, which results in the creation of an abnormally short mRNA transcript. This mRNA strand, when translated, yields an abnormal protein that is permanently affixed to the nuclear rim, and therefore does not become part of the nuclear lamina. As a consequence, the nuclear lamina is unable to provide the nuclear envelope with adequate structural support necessary for the organizing of chromatin to condense into chromosomes during mitosis, which limits the ability of the cell to divide.

Gene mutations can also lead to "spontaneous" insertions or deletions of a number of nucleotides that is not evenly divisible by three from a DNA sequence. Due to the triplet nature of gene expression by codons, the insertion or deletion can change the reading frame (the grouping of the codons), resulting in a translation that is completely different from the original. This is called a "frame-shift" mutation— also called a framing error or a reading frame shift. The earlier in the sequence the deletion or insertion occurs, the more altered the protein will be.

What causes such mutations? In most of the cases, mutations are caused by environmental factors known as *mutagens*. When a mutagen causes a mutation that leads to cancer, the mutagen is also called a carcinogen. One study showed that approximately 90% of all known carcinogens are mutagens. Cancer is fundamentally a disease of failure in regulating cell growth. Somehow a normal cell has been transformed into a cancer cell by mutations in those genes which regulate cell growth and differentiation. The affected genes are divided into two broad categories. Oncogenes are genes which *promote* cell growth and reproduction. Tumor-suppressor genes are genes which *suppress* cell division and survival. Either type can be disrupted by mutations.

The modern environment exposes everyone to a wide variety of chemicals in drugs, cosmetics, food preservatives, plastics, pesticides, pollutants, and so forth. Many of these compounds have been shown to be carcinogenic as well as mutagenic. Examples include the food preservative AF-2, the pesticide ethylene dibromide, the drug hycanthone, several hair-dye additives, and the industrial compound polyvinyl chloride (PVC) used in some plastics. Recently, the National Institutes of Health (NIH) has added formaldehyde and styrene to the list of known and possibly cancer-causing agents; formaldehyde is used to make compounds such as plastics and synthetic fibers; styrene is a synthetic chemical used in the manufacturing of products such as rubber and plastic. The greatest exposure to styrene in the general population, though, is through cigarette smoking.

Another infamous group of mutagens are radioactive substances such as radon, sometimes called radio-nuclides or radioactive isotopes. They are atoms with an unstable nucleus which can cause radiation. Radiation is the emission (sending out) of energetic particles or energetic waves. An important distinction with regard to the health risks from radiation is whether the released energy is ionizing or non-ionizing. Non-ionizing radiation is a low-frequency radiation that does not have enough energy to remove electrons or directly damage DNA. Low-energy UV-rays, visible light, infrared rays, microwaves, and radio waves are all forms of non-ionizing radiation. Aside from UV-rays, these types of radiation are not known to increase mutation risk.

Ionizing radiation, on the other hand, is high-frequency radiation which has enough energy to remove an electron from an atom or molecule, thereby ionizing it. Ionizing radiation has enough energy to damage the DNA in cells, which in turn may lead to cancer. Gamma rays, X-rays, and some high-energy UV-rays are forms of ionizing radiation. The amount of damage in the cell is related to the dose of radiation it receives.

It is important to understand the difference between these two types of radiation. For example, the non-ionizing radiation given off by a cell phone or a television screen is not the same as the ionizing radiation you might get from X-rays taken in the hospital. Ionizing radiation, however, is a proven human carcinogen. The evidence for this comes from many different sources, including studies of atomic bomb survivors in Japan, people exposed during the Chernobyl nuclear accident, people treated with high doses of radiation for cancer and other conditions, and people exposed to high levels of radiation at work, such as uranium miners, or at home (radon). In general, the risk of cancer from this kind of radiation exposure increases as the dose of radiation increases.

In terms of energy, UV-rays straddle the border between ionizing and non-ionizing radiation. They have more energy than visible light, but not as much as X-rays. However, UV-rays often have enough energy to damage a cell's DNA. But because they do not have enough energy to penetrate deeply into the body, their main effect is on the skin. UV-B light causes crosslinking between adjacent cytosine and thymine bases in a DNA string and leads to an abnormal covalent bond between the two bases on the same side of the helix. Such abnormal bonds interfere with base pairing during DNA replication, leading to mutations. This is called direct DNA damage. Then there is UV-A light that can cause indirect DNA damage by creating free radicals which may trigger DNA damage.

There they are—radicals. An important cause of mutations comes from so-called free radicals. These are atoms, molecules, or ions with unpaired electrons. With some exceptions, these unpaired electrons cause radicals to be highly chemically reactive; they are prone to losing or picking up an electron, so that all electrons in the atom or molecule will be paired. Many forms of cancer are thought to be the result of reactions between free radicals and DNA. This process may result in mutations that can adversely affect the cell cycle and potentially lead to cancer. This makes free radicals powerful mutagens, and thus also carcinogens. Laboratory and animal research have shown that antioxidants may help prevent the free radical damage that is associated with cancer—they act like free-radical scavengers. Antioxidants are molecules which can safely interact with free radicals and terminate the chain reaction before vital molecules such as DNA are damaged. Examples of antioxidants include beta-carotene, lycopene, resveratrol, vitamins C, E, and A; they naturally come with a healthy diet that includes a variety of fruits and vegetables.

Let us come to a conclusion. Mutations are the main cause of variability in the gene pool. Every day, researchers around the world report new disease-associated mutations in medical journals. Such studies show that we are all walking genetic "junkyards." Recent research suggests that every individual carries, on average, 313 disease-causing mutations. Fortunately, not all mutations cause diseases.

The changes in DNA caused by mutation can have very different effects. Think of the following analogy: When we drop a watch on the floor, the effects can vary widely—from being destructive (broken) to neutral to constructive (working again). Well, something similar happens after mutation. Some mutations alter a gene's DNA base sequence but, because of redundancy in the DNA code due to codon synonyms, they may not change the function of the protein that the gene produces. Other mutations cause errors in protein sequence, creating partially or completely non-functional proteins, which may be the cause of a disease—but not necessarily so. And then there are mutations that can be beneficial for the organism in a certain environment.

Genetics is basically the study of differences. Human biology has shown us that there is much variation and variability in the human race. When you study anatomy, you might think we all have the same anatomical structure, until you discover that differences abound (some even have an appendix on the "wrong" side). The same story holds for human genetics. When you study the results of the Human Genome Project, you might think we all have the same genetic outfit, until you discover that differences abound. There is no standard genome.

We all know of harmful mutations. But are there also beneficial mutations? Yes, there are. The following list has just a few examples of various mutations that have been advantageous for the human gene pool.

1. All humans have a gene for a protein called apolipo-protein AI, which is part of the system that transports cholesterol through the bloodstream. A small community in Italy is known to have a mutant version of this protein, named apolipo-protein AI-Milano, or Apo-AIM for short. Apo-AIM is even more effective than Apo-AI at removing cholesterol from cells and dissolving arterial plaques, preventing some of the damage from inflammation that normally occurs in arteriosclerosis. Perhaps something similar is the case among the Maasai of Kenya who are heavy meat eaters and milk consumers, yet have low cholesterol levels.

2. One of the genes that govern bone density in human beings is called low-density lipoprotein receptor-related protein 5, or LRP5 for short. Mutations which impair the function of LRP5 are known to cause osteoporosis. But a different kind of mutation can amplify its function, causing one of the most unusual human mutations known. This mutation was first discovered fortuitously, when a young person from a Midwest family was in a serious car crash from which they walked away with no broken bones.

3. Many adults lack a beneficial mutation that enables others to eat milk and dairy products—it is called lactose intolerance and is common in many populations in Eastern and Southeastern Asian, but also among some Africans. Within the past 10,000 years, this mutation became beneficial and had a positive fitness value among those who practice dairy farming. It is caused by a mutation that allows the enzyme lactase to remain expressed after stopping breast feeding. The expression of the lactase gene (LCT, located on chromosome 2) is under control of its promoter which is located just upstream of the lactase gene and facilitates the transcription of the LCT gene. However, there are two mutations or SNPs (single-nucleotide polymorphisms) that appear to have an enhancer effect on the lactase promoter:

T–13910 (T, instead of C, at position -13910 upstream of the gene LCT) and A–22018 (A, instead of G, at position -22018). It turns out that the lactase gene has a higher expression when these mutations are present, but a lower expression when not. Somehow these single point mutations move or remove the binding sites of individual proteins (see Chapter 9).

4. There is also a mutation in a membrane receptor protein that confers resistance to HIV for homozygotes and delays AIDS onset in heterozygotes by preventing HIV from binding to cells. It is most likely a remnant from resistance to the bubonic plague or smallpox. This might explain why this mutation is not found in southern Africa, where the bubonic plague never reached. Because HIV has not been around for more than one or two generations, there has not been long enough selective pressure for it to spread throughout the population.

5. EPAS1 is a gene that codes for a protein involved in responding to a falling oxygen level. It seems to be the key to Tibetan adaptation to life at high altitude where there is 40% less oxygen in the air than at sea level. A mutation in the gene that is thought to affect red blood cell production is present in only 9% of the Han population, but was found in 87% of the Tibetan population.

6. Among insects, resistance to pesticides has been spreading due to a high selective pressure for certain mutations. DDT-resistant mosquitoes, for instance, make an enzymatic protein that can accommodate a single molecule of DDT and inactivate it by adding oxygen to a chlorinated side group on its molecule. Every once in a while, a malaria mosquito must have had a gene mutation that is capable of inactivating DDT—perhaps as a result of a duplicated and mutated gene in the "silent" DNA section (more on this in Chapter 11).

7. Something similar may have happened among bacteria that are resistant to antibiotics. One way of resistance is the production of penicillinases, a group of beta-lactamase enzymes that cleave the beta-

lactam ring of a penicillin molecule such as streptomycin. Since penicillin is a group of antibiotics derived from *Penicillium* fungi, bacteria have always been exposed to it, so some of them may have developed the right DNA code to break penicillin down. The medical use of penicillin on a large scale would be a highly selective pressure for the spread of such advantageous DNA; the spread can go relatively fast because bacteria have a quick life cycle.

In short, mutations are most of the time either neutral or harmful, but they can also be beneficial every once in a while (more cases of beneficial mutations are discussed in Chapter 11).

11. The silent parts of DNA

So far, we have mainly been talking about DNA that is located in genes—perhaps interrupted by introns that will be excised during the transcription process—but there is much more DNA in between those genes, so-called inter-genic DNA. This DNA holds DNA sequences that do not seem to encode protein sequences, and is therefore usually called non-coding DNA. The amount of non-coding DNA varies greatly among species, but in the human genome up to 98% is considered non-coding DNA.

Initially, a large proportion of non-coding DNA had no known biological function and was therefore sometimes referred to as "junk DNA," particularly in the lay press. This is, however, a very misleading term, because it has become more and more evident that this DNA is not as useless as initially thought. This alleged "wasteland" has proved to be a repository for a variety of functions that are part of normal, and even critical, cellular processes. For that reason, it is much better and safer to speak of "non-coding," "neutral," or "silent" DNA. Just because the function of a specific region is not known does not mean that it has no function at all. Absence of evidence is not evidence of absence.

What then could the function of this "silent" DNA be? There are several possibilities. Some sequences may have no biological function for the organism itself, but for some "intruder" such as endogenous retroviruses they do. A retrovirus is an RNA virus that is duplicated in a host cell using the reverse transcriptase enzyme to produce DNA from its RNA genome; the DNA is then incorporated into the host's genome by an integrase enzyme; thereafter the virus replicates as part of the host cell's DNA. This DNA is not functional for the host organism, but it obviously is for the intruder. When a retrovirus is inserted into oncogenes, it can convert normal cells into cancer cells (see Chapter 10).

Then there is "silent" DNA that controls the DNA of "physical" genes. Cis-regulatory elements control the transcription of a nearby gene, whereas trans-regulatory elements control the transcription of a distant gene. This could also be the region where non-coding RNAs are located. Non-coding functional RNAs are functional molecules that are not translated into protein. Though little is known about them, they are thought to have regulatory functions, including the transcriptional and translational regulation of protein-coding sequences. Examples of noncoding RNA include ribosomal RNA (rRNA), transfer RNA (tRNA), and micro-RNA (miRNA). Micro-RNAs are predicted to control the translational activity of approximately 30% of all protein-coding genes in mammals.

Another function of "silent" DNA could be that they dampen the effect of "frame-shift" mutations (see Chapter 10). Since non-coding DNA separates genes from each other with long gaps, a "frame-shift" mutation in one gene or part of a chromosome may not cause a "frame-shift" for the rest of the chromosome, because "silent" DNA could minimize the changes caused by such mutations.

Then there is at least one more function "silent" DNA may have for the organism—redundancy. Non-coding DNA appears to harbor large numbers of tandem, repetitive DNA sequences. These are most likely the result of gene duplications. Let me explain first how gene duplications may occur.

One of the main mechanisms is a so-called *unequal* crossing-over or recombination. When two sequences are misaligned, unequal crossing-over may create a tandem repeat on one chromosome and a deletion on the other. This is a type of gene duplication that inserts and deletes at the same time; it deletes a sequence in one DNA strand and replaces it with a duplication from its sister strand, so one chromosome will lose a DNA sequence whereas the other chromosome of the same pair will gain the DNA sequence lost by the first. Unequal cross-ing-over requires a measure of similarity between the sequences for misalignment to occur. The more similarity within the sequences, the more likely unequal crossing over will occur, allowing for the mismatch in the cross-over point to occur.

Another mechanism that may create duplicates is based on transpos-ons and retro-transposons—which are sometimes called "jumping genes." Transposons can either *cut*-and-paste or *copy*-and-paste; in the latter case they create duplicates; in the former case they can still insert several copies at the target site. Retro-transposons, on the oth-er hand, always copy themselves, but in two stages, first from DNA to RNA by transcription, then from RNA back to DNA by reverse tran-scription. The DNA copy is then inserted into the genome in a new po-sition. Reverse transcription is catalyzed by a reverse transcriptase enzyme, which is sometimes coded by the transposable element itself.

To estimate how many duplicates of various genes a genome carries, geneticists use DNA *micro-array* technology. When large quantities of the DNA from a given genome attach to certain parts of the micro-array chip, this is an indication of how many copies of a specific gene that genome contains. By using this method, it was found, for in-stance, that about 15,000 of the more than 40,000 genes in the hu-man genome appear to have been produced by gene duplication (see also Chapter 2 where we discussed CNVs).

A classic example is the human beta-globin cluster. It is composed of five genes located on a short region of chromosome 11. There is actu-ally also a sixth gene, but it is a pseudo-gene that has a mutation which prevents its expression. The sequences of these six genes are quite similar, which suggests they occurred by duplication of an ances-tral beta-globin gene. Together all five functional genes are responsi-ble for the creation of the beta parts (roughly half) of the oxygen transport protein hemoglobin but they do so at different moments

during development: The epsilon gene is expressed during early embryo development, the two gamma genes during fetal development,
the delta gene early after birth, and the beta gene throughout the remainder of the life cycle.

The order of the genes in the beta-globin cluster is: 5' – epsilon –
gamma-G – gamma-A – delta – beta – 3'. Expression of all these genes
is controlled by a single locus control region (LCR). The arrangement of
the genes directly reflects the temporal differentiation of their expression during development, with the early-embryonic stage version of
the gene located closest to the LCR. Interestingly enough, if the genes
are rearranged, the gene products are expressed at improper stages of
development.

How could the creation of gene duplicates ever be advantageous to
the organism? Well, the main reason seems to be that this redundancy
is a rich source for *new* functional genes. The duplication of a gene
results in an additional copy that is free from selective pressure—that
is, mutations of it have no deleterious effects to its host organism.
Thus it can accumulate mutations faster than a functional single-copy
gene, over generations of organisms. This freedom from consequences allows for the mutation of novel genes that could potentially code
for a new function and increase the fitness of the organism.

Of course, duplicates do not always lead to new functional genes.
There are several ways the duplicate gene can evolve. Sometimes, it
undergoes mutations without much effect. Sometimes the duplicate
copy increases the dosage effect of the gene product, so the duplicate
may be retained as a redundant copy. Another possibility is that the
duplicated copy acquires a beneficial mutation that gives it a function
different from its ancestor. If so, we could have a new gene in the
making! Then there is also the possibility that mutations in the duplicate are harmful. When this happens, the duplicate becomes a pseudo-gene. Yet, pseudo-genes can still undergo mutations that may pro-

duce in time a new beneficial protein. Pseudo-gene sequences appear to accumulate mutations more rapidly than coding sequences due to a loss of selective pressure.

An example of a "proto-gene"—a new gene in the making—is the apparent mutation of a duplicated digestive gene in a family of ice fish which changed the gene into an antifreeze gene. Another example is a pancreatic enzyme, RNASE1, which breaks down bacterial RNA in the intestines. Most primates have one gene encoding the enzyme, but it was found that colobine monkeys from Asia have two versions: One encodes RNASE1, while its duplicate encodes a new enzyme, RNASE1B. The original enzyme works best at pH 7.4, but the new enzyme works six times better than RNASE1 at pH 6.3, which is the acidity level of the colobine monkeys' small intestines. Before the duplication, one enzyme did two jobs, but after duplication, two enzymes do just one job each, but one doing it better than the other.

This phenomenon is most likely also the explanation of how the human blood clotting mechanism came along. It follows an intricate cascading route with a dozen or so proteins; these proteins are also called "factors." Most hemophiliacs carry a mutated version of factor VIII, some of factor IX. Most of these proteins turn out to be related to one another at the level of amino acid sequence, which is an indication of ancient gene duplications, most likely in "silent" DNA sections.

This explanation is even more plausible when we realize that fish have a much shorter cascade of blood-clotting proteins; the longer cascade found in mammals has some extra proteins very similar to the original ones, which makes it highly probable they are a result of replicated DNA sections. Since new copies were not essential for the original function, they could gradually evolve to take on a new function, driven by the force of mutation and natural selection. Our current blood clotting mechanism is able to stop possible leaks much more quickly as it evolved from a low-pressure to a high-pressure cardio-vascular system—which required extra new proteins.

Another case of gene duplication well documented is the evolution of color vision. In a vast majority of vertebrates, there are two different sets of photoreceptors, one that operates during the day and another that operates in the dark. Rods mediate night vision, cones day vision. The first detector of light energy is a chromophore, actually a vitamin A derivative. Each chromophore is bound to a membrane protein called an opsin (which is an apo-protein). It is the DNA sequence of an organism's opsin proteins that determines the spectral sensitivity of its cone cells.

Among vertebrates, the rod opsin seems to be the original one, whereas cone opsins have arisen mainly by duplication and subsequent mutation of the rod opsin gene. Today, most mammals possess two-color vision, corresponding to red-green color blindness. They can thus distinguish between violet, blue, green and yellow, but cannot distinguish ultraviolet, reds, and oranges. This was probably the case for the first mammalian ancestors, which were likely small, nocturnal, and burrowing organisms. Primates later developed three-color vision through cone cells that have spectral peaks in the violet (short wave, S), green (middle wave, M), and yellow-green (long wave, L) wavelengths.

In humans, the S-cone pigment gene is located on chromosome 3, whereas the M- and L-cone pigment gene are positioned on the X-chromosome. Curiously enough, the M- and L-cone pigment genes lie next to each other and are 98% identical at the DNA sequence level, which is a strong indication they were once duplicated of each other. At first these genes would have been identical, but over time they diverged to become distinct cones with separated wavelength sensitivity, thus adding sensitivity to red wavelengths. That would be the beginning of three-color vision.

Since human L and M pigment genes are located next to each other on the X chromosome and are 98% identical, hybrid or chimeric genes are commonly created. Hybrids are fusion genes containing the coding sequences of both L- and M-cone pigment genes. These produce a variety of slightly different pigments that are capable of producing significant spectral shifts. The peak spectral sensitivities of L and M pigments are separated by some 30 nm, and most of this separation is due to amino acid substitutions at two of the seven positions. So hybrids may lead to M-like or L-like pigments. It has been suggested that, as women have two different X chromosomes in their cells, 2–3% of the world's women might have a fourth type of cone that lies between the standard red and green cones, giving them four different simultaneously functioning kinds of cone cells.

It is time to return to the issue of non-coding, neutral, or "silent" DNA. This 98% chunk of DNA turned out not to be the "wasteland" is was once thought to be, but it has several possible functions as we discussed earlier in this chapter. However, what has taken many by surprise is that the non-coding DNA, even beyond its regulatory elements, has conserved DNA sequences from many generations ago. Therefore it can act like a logbook of human history by letting us trace back what the DNA of our first human ancestors must have looked like.

This is done by comparing the "silent" DNA sections of two specific types of DNA: mitochondrial DNA and Y-chromosomal DNA. Why these two? Well, DNA is not only present in the cell nucleus but also in its mitochondria, which are the energy providers of a cell (mtDNA). All mitochondria in the egg cell come from the mother only, since sperm cells have no mitochondria. Since mitochondrial DNA does not take part in recombination, mtDNA in the offspring is identical to that from the mother. In other words, mitochondrial DNA is passed un-mixed from mothers to all her children, along the maternal line. Something similar holds for Y-chromosomal DNA too. Since there is no counterpart to the Y-chromosome in the egg cell, it does not take part in re-combination, so the Y-chromosome in the son is an identical copy of the Y chromosome in the father. It is passed un-mixed from fathers to all his sons, along the paternal line.

The "silent" parts of these DNA sections can be tested for single-nucleotide-polymorphisms (SNPs); these are single base pair changes that occur throughout the genome, including its "silent" DNA sections. Any specific inherited SNP is known as a *haplotype*. It is called haplo-type because it is only present in one chromosome of the pair, that is, in a haploid state. Haplotypes can be in any of the 46 chromosomes, but in genealogical ancestry research, the term usually refers to "si-lent" sections of the Y-chromosome and of mitochondrial DNA. Thus there are Y-chromosome haplotypes, which trace patrilineal ancestry, and mitochondrial haplotypes, which trace matrilineal ancestry.

Since SNPs can last through thousands of generations, they can be used as markers to trace an individual's ancestry. Hence, mutations to non-essential portions of the DNA are useful for measuring time—the so-called molecular clock. It is assumed that such mutations occur with a uniform probability per unit of time in a particular portion of DNA.

If a certain SNP is present in a number of individuals, it came assumed-ly from a common ancestor, as it is very unlikely that the same SNP occurred twice in two unrelated populations. Because of this, geneti-cists do not rely on a single SNP but on a more extensive DNA analysis based on several known SNPs. The end result is like a "DNA signature" of their ancestry.

The math behind this can the simplified as follows. If P is the percent-age of no-mutations in a year in a certain DNA segment, then P^N is the probability of no-mutations over N years. On average, given two indi-viduals who had a common ancestor many generations ago, you would expect that the percentage of segments that are mutated in one or the other is, on average, $2(1 - P^N)$. This is an estimate of the percentage of segments that would be found different when comparing two individuals with a common ancestor *N* years ago.

Based on such calculations, the common ancestor of two individuals who have a 10% difference in their DNA markers, must have lived 100,000 years ago if the mutation rate for those DNA segments is 0,9999995, but 250,000 years ago based on a rate of 0,9999998. Apparently, small differences in mutation rate can have an enormous impact. In other words, the accuracy of the molecular clock depends heavily on the accuracy of the mutation rate.

A group of individuals who share similar haplotypes is known as a haplo-group, identified by certain DNA markers of rare mutations in noncoding DNA segments. All individuals in a given haplo-group have a common ancestor at some point in time. Each haplo-group can then split into subgroups that have some additional markers. If letters represent certain markers in the following simple example, then we could assume that haplotype ABCDE is the ancestor of both haplotype aBCDE and haplotype ABCDe—and together they would form a haplo-group.

If enough individuals are examined for their haplotypes, maps can be created, showing the distribution of various haplo-groups across the world. These maps are called haplotype maps, and they provide a variety of interesting information, such as the history of past migrations of people to different parts of the world. This kind of analysis has allowed us to construct a DNA family tree that has at its origin a "mitochondrial Eve" and a "Y-chromosomal Adam" representing our most recent common ancestors that have DNA markers like ours, apparently going back to a location somewhere in Africa.

From the calculations explained above, it appears that "Y-chromosomal Adam" lived about 60,000-90,000 years ago in Africa. Again, the date is rather approximate, since such calculations are not exact. In a similar way, everyone alive today can also be linked back to "mitochondrial Eve." She lived about 140,000-120,000 years ago, so she was not contemporary with "Y-chromosomal Adam," who lived much later.

This may seem odd, but a comparison might help here. If a disaster would wipe out humanity except one man—let us call him Noah—plus his wife, his sons, and their wives, these people would be the ancestors of all descendants coming forth from them. Now this Noah would be the most recent common male ancestor through a strictly male lineage—he would be the "Y-chromosome Adam." But "mitochondrial Eve" would not be Noah's wife, but the most recent common female ancestor of his daughters-in-law through a strictly female lineage. This female ancestor must go back much earlier than Noah himself.

Please be aware that "mitochondrial Eve" and "Y-chromosomal Adam" are not concrete individuals but simply mathematical concepts—the most recent common matrilineal and patrilineal ancestors. Certainly do not confuse them with Adam and Eve as mentioned in the Bible. Genetics studies the *Book of Nature* in which "Y-chromosomal Adam" and "mitochondrial Eve" represent a branching point in the tree of evolution, whereas the Judeo-Christian religion studies the *Book of Scripture*, in which Adam and Eve represent the "grandparents" of humanity who started the first dysfunctional family. We cannot identify or equate them; we cannot read the *Book of Scripture* as if it were the *Book of Nature*, nor reversed. The *Book of Nature* tells us where we come from in a biological sense, whereas the *Book of Scripture* tells us where we come from in a religious sense. These two books complement each other, as they have the same Author—a match made in Heaven.

12. DNA, the Secret of Life?

After all we have seen so far, DNA only seems to have become more powerful than ever. Nevertheless, I would rather argue again that DNA may be everywhere, but DNA is not all there is. I would stress the point that it is not all in the genes, let alone in the DNA. Let me spend the remainder of this book to underpin this pivotal point that puts genetics back in its right place.

I think the whole question boils down to this: Does the "secret of life" really reside in DNA? Is DNA all there is and all that matters in the course of a human life? I would say the answer to these questions is a definite no. Why?

First of all, DNA can never do anything on its own. It is not even capable, as many still believe, of self-replication—that is, making copies of itself on its own—for DNA is manufactured out of small molecular bits and pieces by the use of an elaborate cell machinery that is made up of proteins. If DNA is put in the presence of all the pieces that will be assembled into new DNA, but without any protein machinery, nothing happens. It is actually the presence of many other components that makes sure old DNA strands are replicated into new strands.

This process is analogous to the production of copies of a document by an office copy machine—a process that would never be described as "self-replication." Think of viruses, which are essentially pure DNA or RNA; their DNA or RNA cannot do anything until they penetrate, like a Trojan horse, the interior of a "living" cell where they utilize the cell machinery. So we have a chicken-and-egg problem here: DNA requires proteins, and proteins require DNA.

Second, DNA on its own does not produce anything, not even proteins! The role of DNA is to provide a specification as to how amino acids are to be strung together into proteins by some synthetic machinery, but this string of amino acids is not yet a protein. To become a protein with physiological and structural functions, certain parts may have to be excised and then the string must be folded into a three-dimensional configuration that is only partly based on its amino acid sequence, but is also determined by the cellular environment and by special processing proteins.

Insulin for diabetics makes a case in point. Recently, the DNA coding sequence for human insulin was inserted into bacteria, which were then grown in large fermenters until a protein with the amino acid sequence of human insulin could be extracted. But it turned out amino acid sequence alone does not determine the shape of a protein. The first proteins harvested through this process did have the correct sequence, but were physiologically inactive. Imagine what had happened: The bacterial cell had folded the protein incorrectly! Somehow the DNA for a protein does not "know" how to fold a protein, so as to make it work. This may happen more often than initially thought. Amyloids, for instance, are insoluble protein aggregates that arise from inappropriately folded polypeptides naturally present in the body. These wrongly folded structures may even play a role in various neurodegenerative disorders, including Alzheimer's disease. In other words, DNA surely does not know everything!

A similar situation may also explain how prions cause diseases such as Creutzfeldt-Jakob disease. Prions propagate by transmitting a protein in a wrongly folded state. When a prion enters a healthy organism, it induces existing, properly folded proteins to convert into the disease-associated prion form. Somehow the prion acts as a template to guide the folding process of more proteins into prion form. These newly formed prions can then go on to convert more proteins themselves; this triggers a chain reaction that produces large amounts of the prion form. This process is only possible because DNA does not quite "know" how to fold proteins.

Third, the proteins that DNA produces may not only need a special folding process but they are also incomplete yet—they may require the presence of additional, non-protein factors that are not under direct DNA control, but come from the environment. Many proteins, especially those with enzymatic activity, need "helper molecules" to perform their biological function. These factors can be loosely-bound, so-called coenzymes, or tightly-bound, so-called prosthetic groups.

The enzymes alcohol dehydrogenase and DNA polymerase, for instance, require the presence of zinc as a cofactor for them to work. The hemoglobin protein requires the presence of a prosthetic heme group that contains iron in its center. The most common cofactors are metal ions such as iron, zinc, and copper; other cofactors are vitamins (vitamin C) or are made from vitamins (B-vitamins). Without these cofactors, many DNA products cannot function properly, so in such cases DNA delivered merely an unfinished, non-working product.

Fourth, there are heritable changes in gene activity which are not caused by changes in the DNA sequence, but by mechanisms such as DNA methylation and histone modification, each of which alters how genes are expressed without altering the underlying DNA sequence. The study of such phenomena is called epigenetics.

DNA methylation is a biochemical process whereby a methyl group is added to the cytosine or adenine DNA nucleotides. DNA methylation may affect the transcription of genes in two ways. First, the methylation of DNA itself may physically impede the binding of transcriptional proteins to the gene, and second, and likely more important, methylated DNA may be bound by proteins known as methyl-CpG-binding domain proteins (MBDs). DNA methylation can stably alter the expression of genes.

Histone modifications act in diverse biological processes such as gene regulation, DNA repair, chromosome condensation (mitosis) and spermatogenesis (meiosis). They can activate or rather repress genes.

Histones are highly alkaline proteins found in eukaryotic cell nuclei that package and order the DNA into structural units called nucleosomes. This enables the compaction necessary to fit the large genomes of eukaryotes inside cell nuclei: the compacted molecule is 40,000 times shorter than an unpacked molecule. Histones are the chief protein components of chromatin, acting as spools around which DNA winds, and as such they can play a role in gene regulation.

Fifth, I would like to bring in a more philosophical argument. If human beings were really nothing more than DNA, then it must have been the DNA of two scientists—the ones who discovered DNA, Watson and Crick—that discovered DNA. Think about that statement for a moment—DNA being discovered by DNA. If Watson and Crick were really nothing but DNA, then Watson's and Crick's DNA must have discovered itself. That would be real magic. That would be like the magic of a projector projecting itself or a copy machine copying itself.

We all know that we cannot pull ourselves up by our own bootstraps. Watson and Crick must have been more than their DNA in order for them to be able to discover DNA. If we did not realize this, then we could foolishly demote organisms, including human beings, to merely being DNA's way of creating more DNA—a form of "selfish DNA," so to speak. Science-fiction is around the corner.

Sixth, if we were nothing but DNA, this very statement—claiming that we are nothing but DNA—would not be worth more than its molecular origin, and neither would we ourselves who are making such a statement. Claims like these just defeat and destroy themselves. They cut off the very branch that the person who makes such claims is—or actually was—sitting on.

If we want to accept the reliability of our biological knowledge regarding DNA, we cannot conclude at the same time that all human knowledge is just a product of DNA. That would be "irrational" suicide! In other words, those who make such claims must be "more" than the

DNA they carry. You are always more than your DNA code, for you could have an identical twin, with the same genetic code, but you are you, never him or her.

Seventh, there is this problem as to where DNA comes from. Sure, it comes from your parents, and they in turn received it from their parents. But that is not what I mean. Where did it ultimately come from? Most people will say it originated from the animal world through a process of evolution based on natural selection. I do not want to go into the discussion as to whether this is true, or whether this is even possible (I did so in other books). I have come to the conclusion, though, that it is basically true. But even if it is true, my question would be: How can DNA have the capacity to act as a "coding" language? Where does this capability ultimately come from?

I see no other answer than this: Our universe was created with an underlying *cosmic design* that regulates what is and what is not possible in this world. All our laws of nature are part of this cosmic design. The chemical element carbon (C), for instance, has the "built-in" capability to form very long chains of interconnecting C-C bonds, which allows carbon to form an almost infinite number of compounds. That is the reason why our entire organic chemistry is carbon-based, making carbon the "favorite" building block of the living world on planet earth. So it should not surprise us that carbon is an essential part of DNA as well—that is the way our world was designed. When a die constantly throws a six, I would say it must be loaded. Well, our world seems to be "loaded" too—loaded with cosmic design.

It seems to me that the above considerations have taken DNA off its acclaimed pedestal. This gives me reason to make the daring statement that DNA is not so much the "secret of life," as life is the "secret of DNA." Those who say "it's all about DNA in life" belong actually to the same breed of people as those who shout "it's all about money" or "it's all about sex," or "it's all about politics." Those who narrow their outlook on life to statements like these no longer realize that there is so much more to life. Such seemingly broad-minded slogans are actually narrow-minded ideologies which easily contradict each other. Fortunately most of us know how to take such expressions, as long as we do not take them literally.

Once we realize that DNA cannot do anything on its own—because it needs proteins, for one thing—we might wonder where DNA came from in the evolution of life. What was first—DNA or proteins? DNA makes proteins but it needs proteins for it to work. This is essentially a chicken-or-egg problem. Perhaps there is a solution to this problem.

The origin of life took place some 4.5 billion years ago. Although there was obviously no reproduction before life existed, neo-Darwinism assumes that we can apply something like "natural selection" to the very origin of life as well. I will briefly explain the theory of the Belgian biochemist and Nobel laureate Christian De Duve. According to De Duve's reconstruction, emerging life went through four main successive stages—the primeval pre-biotic world, the thio-ester world, the RNA world, and the DNA world. His version is very definite as to the part played by the thio-ester world, which may be the solution for our chicken-or-egg problem. Let us only discuss some central points.

Stage 1 was characterized by an abundant source of energy (ultraviolet light) and an abundant source of electrons (coming from Fe^{2+} ions, which are essential on a planet that was still lacking molecular hydrogen). Thanks to this combination, the atmosphere became enriched with molecules such as H_2, CH_4, NH_3, HCN, H_2O, and H_2S. Through synthetic reactions sparked by electric discharges and other physical forms of energy, these were synthesized into small organic molecules such as carboxylic acids, including amino acids, hydroxyl acids, and thiols (based on H_2S; coenzyme A, for one, is a thiol). These substances accumulated in oceans and lakes as a "prebiotic broth." In the laboratory, such synthetic reactions were repeatedly confirmed.

Stage 2, the thio-ester world, is deemed necessary because the "prebiotic broth" would have remained "sterile" without additional help of both catalytic and energetic nature. This step was accomplished by thio-esters arising from the condensation of carboxylic acids with thiols. Thio-esters are capable of linking amino acids together into proteins, thanks to energy derived from the consumed thio-ester bonds. The small subset of polymers that "survived" in this process included a number of crude catalysts fore-shadowing present-day enzymes. Is this real natural selection? Not in a strict sense, of course; call it "molecular selection" if you wish.

Why are these "enzymes" so important? Well, a modern cell can only produce proteins with the help of DNA, RNA, and enzymatic proteins. The making of proteins is currently based on DNA and RNA, but DNA and RNA can only achieve this by using enzymatic proteins. So that is where we run into our chicken-or-egg problem: no DNA and RNA without proteins, and no proteins without DNA and RNA. But the thio-ester world would solve this problem, because thio-esters are capable of linking amino acids together into proteins. In addition, scientists have discovered some special RNA molecules (ribozymes) which can also function as enzymes.

Stage 3, the RNA world, developed a small set of proto-tRNAs. Molecules that possessed a site capable of specifically binding a given kind of amino acids were particularly favored by "molecular selection." Reciprocally, the amino acids that were recognized by proto-tRNAs were themselves selected to be used in a metabolism that depended on RNA binding. The central process of protein synthesis was thus born— but essentially with the help of thio-esters.

Stage 4, the DNA world, went through a relatively simple transition which required only a set of enzymes for reverse transcribing of DNA back into RNA. Except for RNA viruses, believed to be vestiges of the

RNA world, RNA replicase and reverse transcriptase enzymes have largely disappeared from the living world. But once the transition was made, the scene was set for the first cells—some 3.5 billion years ago. The first cells resembled present-day bacteria in their structure and their main metabolic attributes (prokaryotes). They occupied hot and acidic sulfurous waters; they were anaerobic in their metabolism, for there was no free oxygen yet. All the forms of life that compose the biosphere today are believed to have arisen from the further evolution of this primeval cell population.

There is at least one more step left: How did prokaryotic cells, which have no mitochondria (the "batteries" of a cell), develop into eukaryotic cells that do possess mitochondria? There is more and more evidence that cell organelles of eukaryotic cells may have originated as free-living bacteria that were incorporated inside a cell. Interestingly enough, mitochondria still contain their own DNA, mtDNA, separate from the DNA found in the cell nucleus. These so-called endo-symbionts could even have transferred some of their own DNA to the host cell's nucleus, since we know that transfer of DNA does occur between bacteria species. Let us leave it at that.

IV. What No Gene Can Do

13. No Gene for Free Will

As I said earlier, DNA may be everywhere, but DNA is not all there is. In this section I am going to explain that we are able to do many things in life that no gene could ever do for us. Is this still science? Yes and no! In the strict sense, it is not science, but more of a philosophical reflection on science (after all, I am someone who also earned a doctorate in the philosophy of science, and I want you to benefit from this). Yet, such a reflection might have huge implications for science.

This discussion probably starts the most controversial section of this book, especially in the eyes of hard-core geneticists who claim that *all* human characteristics can be explained by genetics as has supposedly been done already for *many* of them. However, there is no logically valid way of reasoning from some, or even many, cases to all cases. This is called generalizing induction, which takes us from a general statement about "many" cases to a universal statement about "all" instances, but it is not a logically safe or validated claim. It is at best a belief or conviction. It is a conclusion about all instances of a phenomenon based on some or many instances of that phenomenon. Statements about "many cases" can never be extended to "all cases" in a secure logical way. Apparently, logic can help us to see we are dealing here with claims of a dubious kind. Inductive reasoning cannot even be justified by citing that induction has already been successful in so many cases, for that would make for another inductive argument.

Perhaps philosophy and logic can even be more helpful when it comes to science. I think there is at least one way: Sometimes the tools of logic and philosophy can tell us ahead of time—*a priori*, if you will—what science can and what it cannot achieve. Let me explain this with the following case that I borrowed from the Nobel Laureate and physiologist Peter Medawar.

Anyone looking at paintings of the famous Renaissance painter El Greco will notice that most of his figures are unnaturally tall and thin. Some scientists were eager to explain this in scientific terms. One of them came up with the hypothesis that El Greco must have had a form of astigmatism that distorted his vision and led to elongated images forming on his retina. Sounds interesting, but this hypothesis is doomed from the very beginning. Even if El Greco did see the world through a distorting lens, the same distortion would apply to what he saw on his canvas. These two distortions would cancel each other out, and the proportions in pictures would remain realistic. So we must come to the conclusion that El Greco's figures, particularly the holy ones, appear unnaturally thin and tall because that was his intention, not forced by any gene—he had painted them that way on *purpose*.

Imagine that these scientists had spent lots of time, energy, and funding on testing their hypothesis, not realizing that it was bound to fail ahead of time on purely philosophical, or at least logical, grounds. Yet, some did not give up that easily and tried to attribute Vincent Van Gogh's preference for yellow colors to a visual disorder called xanthopsia; others mentioned drug use or glaucoma. Even if Van Gogh did view the world through a yellow filter, he would also view the colors on his canvas that way. Instead, he must have just chosen the yellow color on purpose!

In all such cases, science just fails us, since philosophy and logic are able to show us *a priori* that the above scientific explanations are doomed to fail, even before any scientific test has been done. There are certain "scientific" ideas you should not spend your time and energy on, because philosophy and logic can show you that they are headed for a dead end. They are inventions that could not possibly make it to discoveries, purely on logical or philosophical grounds.

Something similar might be the case when it comes to issues such as free will, rationality, morality, and religion. All scientific endeavors to anchor these features in genes or DNA might be doomed to fail ahead of time for mere philosophical or logical reasons—as I am going to show you. I admit if science does not go to its limits, it is a failure, but as soon as science oversteps its own capabilities, it becomes arrogant—a know-it-all. Sometimes we know ahead of time—without doing any genetic research—that, in cases like these, a genetic explanation does not make sense or is not even possible. Let me first demonstrate this for what is usually called the "free will" of human beings.

People who believe in strict genetic determinism—and a number of geneticists do, albeit a minority probably—are of the opinion that what some call their "free will" must, like anything else, be determined by genes and DNA. Here are two of their powerful voices. Sidney Brenner, one of the DNA pioneers, said not too long ago he could compute an entire organism, humans included, if he were given its DNA sequence and a large enough computer. With a like mind, the American molecular biologist Walter Gilbert had the audacity to claim that "when we have the complete sequence of the human genome we will know what it is to be human." That is genetic determinism in full glory! Is this the end of our free will? I would not give in too soon...

What is wrong with these statements? Not only is genetic determinism a rather unscientific theory—it steps outside science to claim that there is nothing outside science, which is a claim that could never be experimentally or scientifically tested—but also is it a rather questionable claim from a philosophical and logical point of view. One of its problems is that is leads us into logical trouble. This is one of those cases where we know ahead of time that physical explanations do not make sense or are not even possible. Let me explain.

If we have a free will—and I believe we do as I will argue later on—you might wonder whether this free will could perhaps be based on a gene or a series of genes. Theoretically, there could indeed be a gene that allows us to make choices, but if this gene or additional genes would also determine the outcome of these choices, then we cannot really make free choices and have basically lost the free will we thought we had. Is it possible to eliminate our free will through genetics? Or does this lead us into a *contradictio in terminis*? Philosophy and logic may help us to answer this question ahead of time, "a priori," without us having to do any research in genetics. They tell us that a radical form of genetic determinism would cause philosophical trouble. There are at least three arguments that plead against universal determinism.

The first argument is that a universal form of determinism would lead us into a vicious circle of infinite regress—with no way to get in or out. Here is how. Determinism posits an infinite regress of causes—cause *n* is caused by cause *n-1*, and so on backwards in time. Say, quantum particles are determined, but determined by what? Smaller particles? What determines these particles? Even smaller particles? The chain of causes would have to regress infinitely—which is in itself problematic because an infinite number of things cannot actually exist in a material world.

A second argument goes as follows. Universal determinism creates a loop that makes for a logical paradox caused by the problem of self-reference. Self-reference is used to denote a statement that refers to itself. The most famous example of a self-referential sentence is the liar sentence: "This sentence is not true." If we assume the sentence to be true, then what it states must be false. If, on the other hand, we assume it to be false, then what it states is actually true. In either case we are led to a contradiction. Trying to convince someone of the truth of universal determinism smacks also of self-refutation. How could this claim change someone's mind if everything is fully predetermined anyway? Ironically enough, people who defend universal determinism— let us call them determinists—are willing to spend their entire career on forcing us to choose their conviction that human beings cannot choose.

The third—and in my opinion the strongest—argument is that the world view of universal determinism makes for a contradictory claim. If genes, for instance, really determine everything in my life, then they would also determine my choice to believe or not to believe that genes determine everything. The key problem is that we are dealing here with *beliefs*, and beliefs are not material entities like genes— unlike genes, they can be true or false. If I believe that genes determine everything, I have no reason to suppose my belief is true, and hence I have no reason for supposing that genes determine everything. This is a "boomerang theory" in optima forma—it defeats itself, for once we consider it to be true, it becomes false.

As a matter of fact, beliefs belong to the immaterial world of *thoughts*, and thoughts can be true or false—which sets them apart from all the material things we encounter in this world. In a world of molecules, there is only talk of being small, heavy, strong, and what have you— but not of being true and false, or right and wrong. Suddenly, we find ourselves in a non-material world where things are not large or small, light or heavy, hard or soft—but true or false, and right or wrong. Those who deny the existence of anything immaterial thereby also deny the existence of their very own denial, for all statements, including denials, are immaterial. The world of facts, laws, and beliefs is an immaterial world—perhaps dealing with and referring to the material world, but definitely distinct from the material world

Some might object that thoughts are not immaterial but material entities, like everything else in this universe. Some scientists believe that thoughts are merely neural phenomena based on chemical neurotransmitters and electric impulses. If that were so, this belief would certainly take their immaterial shine away. However, these scientists are claiming something that we could again defuse *a priori* by just lying on our couch and using a philosophical and logical analysis. Let me try to do this by demonstrating that thoughts cannot be mere neural phenomena—or, put in more general terms, that the mind (with its thoughts) cannot be identical to the brain (with its brain waves).

To make their materialistic claim, these scientists love to use the computer analogy: The brain supposedly works in the same way as a computer operates, since both use a binary code based on "ones" (1) and "zeros" (0); neurons either do (1) or do not (0) fire an electric impulse—in the same way as transistors either do (1) or do not (0) conduct an electric current. This makes it look like the brain "thinks" the way a computer "thinks"—it is supposedly the same kind of operation.

But not too fast! There is something wrong here! Whatever is going on in the brain—say, some particular thought—may have a material substrate that works like a binary code, but it would not really matter whether this material substrate works with impulses, as in the brain, or with currents, as in a computer, for the simple reason that this material substrate only acts as a physical carrier for something immaterial—thoughts. Take for instance the speech center that Paul Broca discovered in 1861 in the frontal lobe of the human brain's left hemisphere, near the part of the motor cortex controlling the face and mouth. He found it damaged in patients with aphasia, the inability to produce spoken language. Should we claim that this is a center that creates speech, or should we rather consider it a brain region that merely communicates speech? And what about people with the deaf-equivalent of Broca's aphasia who use sign language? Do they have damage somewhere near the cortex controlling the movement of the hands?

Once we acknowledge that the same thought can be transported by different vehicles—such as pen strokes on paper, sign language, currents in computers, or impulses in the brain—we are logically forced to acknowledge that an immaterial thought is not identical to its material carrier. If I were to break my radio, the news report would stop, but this does not mean the news was created by the radio; it was only the news vehicle that broke down. So it seems evident to me that the brain does not create thoughts but merely transports them. The thoughts coming from the mind somehow "use" the vehicle of the brain; they merely use the machinery of the brain as a tool.

Apparently, there is something enigmatic about the human mind. Let me explain this further with an observation that was made by the philosopher Ludwig Wittgenstein. Picture yourself watching through a mirror how a scientist is studying your opened skull for "brain waves." Wittgenstein once noted correctly that the scientist is observing just one thing—outer brain activities, but the "brain-owner" is actually observing two things—the outer brain activities via the mirror as well as the inner thought processes that no one else has access to. In order for them to make the connection between "inner" mental states and "outer" neural states, scientists would depend on information that only the "brain-owner" can provide.

The world of the mind is only accessible to the "brain-owner." This is even so in court, in spite of lie detector tests; very often the only ones to know whether they committed the crime or not are the defenders themselves. Apparently, there is no mind-reading through brain scans. Would it not be nice if the "brain-observer" could read off from the lighted-up areas of the brain all the knowledge the "brain-owner" had gained in a life time? If this were so and we would still be able to scan Einstein's brain, we could all become little Einsteins. A scanner that could show us the intricate patterns of neurons firing in real time would be a scientific marvel, but viewing the output would bring us no closer to understanding the experience of awareness, the meaning of the thoughts, of what it is like to be that person whose brain is being scanned. Material explanations cannot possibly lead to a full understanding of non-material phenomena.

To sum up, the *brain* is only the material carrier of immaterial thoughts in the *mind*. It is not the brain that does the thinking, but I do—that is, my mind. To use an analogy, it is not the wing that flies, but the bird. I am not trying to *set* brain and mind apart here, but at least I would like to *tell* them apart. We have been brain-washed, though, to think that brain and mind are the same—but if they were, we would run into trouble. Whereas the brain as a material entity has characteristics such as length, width, height, and weight, the mind does not have any of those; thoughts coming from the mind are true or false, right or wrong, but never tall or short, heavy or light (unless taken in a figurative sense).

Therefore, we need to stress that mental activities are very different from neural activities. We cannot just deny the mental, because denying the existence of mental activities is in itself a mental activity, and thus would lead to contradiction. Some people such as the biologist J. B. S. Haldane and the philosopher C. S. Lewis have worded this paradox along the following lines: *If thoughts are nothing but the motions of atoms in the brain, we have no reason to suppose that our beliefs are true ... and hence we have no reason for supposing our thoughts to be composed of atoms.*

I conclude from this that genes could never determine what is true and what is false. There must be something else that determines this. Ultimately, it is our free will that makes such decisions, although our decisions should always be based on the way the world around us actually *is*—which we can find out through observation and experiment as well as through logic and reasoning. Is this not what science is all about? Science tries to tell us scientific, immaterial truths about the material world. People who have unscientific, or just incorrect, thoughts about this world inevitably encounter trouble all the time, because their thoughts do not match reality.

So our thoughts—including those of scientists—must be more than neural activities, otherwise they would be as fragile as the molecules they supposedly came from. That would surely be the end of science—and of all the genetic issues we have been talking about in this book. If the science of genes were the mere product of genes, it would inevitably be a self-defeating enterprise. So let us avoid that cliff!

In other words, there must be something more than genes to regulate our behavior, and our thoughts in particular. Whereas impulses come from the brain and from all that comes with brains—genes, hormones, and the like—thoughts come from the mind. When humans commit a crime, for instance, we tend to look for *motives* in the minds of suspects, not for defects in their brains, let alone in their genes. As human beings, we live in *two* worlds: We are driven by physical causes (hormones, reflexes, drives, genes) as well as by mental causes (motives, reasons, intentions, beliefs). But these two categories harbor very different entities, for mental causes are not under the control of physical causes.

Can we demonstrate the difference between neural events and mental events? To put it in an image: Can the mind change your brain? Most people tend to think the opposite; they believe that a machine—the machinery of your brain, for instance—can change your mind. I think the case is well illustrated by the late neurosurgeon Wilder Penfield when he asked his patient to try and resist the movement of the patient's left arm that he was about to make move by stimulating the motor cortex in the right hemisphere of the brain. The patient grabbed his left arm with his right hand, attempting to restrict the movement that was to be induced by a surgical stimulation of the right brain. As Penfield said, "Behind the brain action of one hemisphere was the patient's mind. Behind the action of the other hemisphere was the electrode." In other words, one action had a physical cause, whereas the other action had a mental cause. That is the enigma of the mind and its free will in a nutshell. Somehow the free will can be a powerful meta-physical force in a physical body. This allows us, to a certain extent, to be masters of our own actions.

Do not ask me *how* body and mind interact—all I try to argue is *that*

they interact. This is not as strange as it may sound. Think of the relationship between mass and gravity; we do not really know how these
two interact. Or take the case of electrically charged particles that interact with each other through the mediation of electromagnetic
fields; the charged particles affect the fields and the fields affect the
particles, but we do not know anything about the "mechanism" behind this interaction. Something similar happens when the free will
interacts with the body—we know *that* they interact but we do not
know *how*.

As Penfield's observation shows us, we can certainly be masters of our
own actions. In the midst of this cause-and-effect world, our free will
can become a mental cause that can have immaterial as well as material effects all by itself. I always like to use the following analogy. When
watching a game on the golf course or on the pool table, we see balls
following precisely determined courses of cause and effect; they follow physical laws and are subject to well-known rules and constraints.
Yet there is one element that does not seem to fit in this predetermined picture, in this precisely determined cascade of causes
and effects—the players of the game themselves. They may work with
a physical model in their minds but they themselves are not part of
that model. Whenever actors act, their acts become causes with their
own effects.

That is why the direction of something like a billiard ball on the pool
table or a golf ball on the golf course is not only ruled by physical laws
but also by human intentions—by players who have a certain goal in
mind. Those very intentions and free decisions are real and do have
consequences; they can become causes that are not part of the physical model but may have physical or non-physical effects of their own.
People who cannot look beyond these physical laws and causes are
completely missing out on what the game is all about. These players
somehow fall outside the realm of the model of physics; they themselves can steer the course of the laws of nature. Every free choice can

become a cause with effects.

Perhaps I should mention one caveat. All free choices have to work within existing boundary conditions—since we cannot effectively choose the impossible. So our freedom of self-determination does not let us do whatever we want to do. The more we are aware of our constraints, the more we can actually be free. "Know yourself," says an old inscription in Delphi. With the proper knowledge, we can take charge of our constraints so that we are no longer their victims, but rather their architects. That is why I said earlier that we can, *to a certain extent*, be masters of our own actions—in spite of what some geneticists seem to suggest.

This issue of our free will becomes even more pressing when it comes to addictions. If we do have a free will, how then is it possible that so many people are struggling with addictions that they seem to be unable to break or stop? Many in our society are addicted to alcohol, to nicotine, to drugs, to overeating, to compulsive shopping, to gambling, to pornography, to endless cell-phone use, to unending video games— the list could go on and on.

In general, we distinguish two kinds of addictions: substance addictions (alcohol, drugs) and behavioral addictions (gambling, sex). Nonetheless, what all addictions have in common is that they go after some form of pleasure. Pleasure is what distinguishes an addiction from an obsessive-compulsive disorder. While people who have addictions suffer all kinds of discomforts, the desire to use a specific substance or engage in a specific behavior is based on the expectation that it will be pleasurable. In contrast, people who experience a compulsion as part of obsessive-compulsive disorder may not get any pleasure from the behavior they carry out.

No doubt, pleasure plays an important role in our lives. If eating food and having sex did not stimulate feelings of pleasure and reward, we would die soon, by lack of food, and we would die out, by lack of sex—natural selection took care of that. These are naturally rewarding behaviors. So in a sense, everyone has some kind of addiction. It is estimated that at least 90% of Americans have at least one form of "soft" addiction in their lives. While it is healthy to relieve stress with behaviors like having a drink and watching television, when they become habitual they also become problematic to one's health and happiness.

That is where the "real" addictions come in—although the borderline is often fuzzy and there may be a sliding slope. How-ever, in general, habits and patterns associated with addictions are usually characterized by short-term rewards—some instant pleasure—coupled with long-term costs—some delayed dam-aging effects. Our brains are wired to ensure that we will repeat life-sustaining activities by associating those activities with pleasure or reward. It is the limbic system in the brain that contains the brain's reward circuit—it links together a number of brain structures that control and regulate our ability to feel pleasure. The limbic system is activated when we perform such activities. It has a dopamine-rich area, which is an intersection where all addictive behaviors meet. Whenever this reward circuit is activated, the brain senses something "great" is happening that needs to be remembered, and motivates us to do it again and again, without thinking about it.

Let us discuss first what the mechanism is behind substance addiction. Reasons for taking drugs or other addictive substances are various—to feel good, to feel better, to do better, to do what others do, to find out what others do out of curiosity—but the ultimate reward is always pleasure, at least in some form and at least in the short term. Most abuse substances directly or indirectly target the brain's reward system by flooding the circuit with dopamine. Dopamine is a neurotransmitter present in regions of the brain that are associated with movement, emotion, cognition, motivation, and feelings of pleasure.

An overstimulation of this system produces the euphoric effects sought by people who abuse drugs and stimulates them to repeat the behavior. These drugs can release 2 to 10 times the amount of dopamine that natural rewards do. In some cases, this occurs almost immediately (as when drugs are smoked or injected), and the effects can last much longer than those produced by natural rewards. The resulting effects on the brain's pleasure circuit dwarf those produced by naturally rewarding behaviors such as eating a regular meal. The effect of such a powerful reward strongly motivates people to take drugs again and again. This is why substance abusers eventually feel flat, lifeless, and depressed, being unable to enjoy things that previously brought them pleasure. Now, they need to take more drugs just to try and bring their dopamine function back up to normal.

There is a similar story for alcohol addiction, better known as alcoholism—although the habituation process may take a bit longer than for drug addiction. The driving force is pleasure again, at least in the short run. When we drink alcohol, most of the ethanol in the body is broken down in the liver by an enzyme called alcohol dehydrogenase (ADH), which transforms ethanol into a toxic, unpleasant compound called acetaldehyde (CH_3CHO), also a known carcinogen. However, acetaldehyde is generally short-lived; it is quickly broken down to a less toxic compound called acetate (CH_3COO^-) by another enzyme called aldehyde dehydrogenase (ALDH). Acetate is finally broken down to carbon dioxide (CO_2) and water (H_2O), mainly in tissues other than the liver.

Regardless of how much a person consumes, the body can only metabolize a certain amount of alcohol every hour. That amount varies widely among individuals and depends on a range of factors, including liver size and body mass, but also on different alleles that cause differences in ADH and ALDH enzymes—another example of genetic polymorphism. Some of these enzyme variants work either more or less efficiently than others. A fast ADH enzyme or a slow ALDH enzyme can cause toxic acetaldehyde to build up in the body, creating dangerous and unpleasant effects.

The story is very similar for other addictive substances such as food, nicotine, caffeine, and chocolate. When analyzing the ad-diction to food for example, researchers have found that the same molecular mechanisms that steer people into drug addiction are behind the drive to overeat, thus pushing people into obesity. They found a particular receptor in the brain known to play an important role in vulnerability to drug addiction—the dopamine D2 receptor. The D2 receptor responds to dopamine, which is re-leased in the brain by pleasurable experiences after consuming certain foods.

The story is not much different for behavioral addictions such as relentless overindulging, extensive gambling, porn addiction, and addiction to the internet or cellphones or video games. There are many similarities in the neurobiology of substance and behavioral addictions. They are both dependent on reinforcement and on reward-based learning processes. Several structures of the brain are important in the conditioning process of behavior addiction. One of the major areas of study includes the region called the amygdala, which is part of the limbic system and involves emotional significance and associated learning.

Let us go back to our original question: If we do have a free will, how is it possible that so many people are struggling with addictions that they seem to be unable to break or stop? Is there a genetic cause that makes them head for a particular addiction? In spite of claims to the contrary, there is no clear evidence that there are genes that force us to become addicted. There may be genes, for instance, that affect alcohol metabolism, as we found out, but that does not mean they also steer alcohol consumption. All they do is giving some people unpleasant experiences after only one or two drinks, which may stop them from drinking sooner than others.

In general, I would say, there are no genes for addictions. Genes may regulate how we *react* to addictions, but it is very doubtful whether they directly *cause* addictions. Alcohol consumption, for instance, has increased dramatically in Western societies recently, making it very unlikely that this is a genetic phenomenon; for nicotine addiction, the opposite is happening due to extensive warning campaigns. No one would assume those are genetic changes over such a short period of time. Here we have a nature-nurture issue again (see Chapter 4). To use an analogy, I do not believe there is a gene for "electro-shock avoidance" (I just invented that one). We try to avoid electro-shocks because they are very unpleasant. Call that inborn, if you want, but that is where genetics ends.

My conclusion is that the initial decision to engage in any kind of sub-stance addiction or behavioral addiction is mostly *voluntary*—a free choice—at least it starts that way. There may be factors, though, that veer the decision in the wrong direction—such as stress, depression, a low self-esteem, or a need to fit in with a group of friends. Yet, the decision itself is an essential part of human freedom, called self-control. Unfortunately, if the decision goes the wrong way, this initial decision can have very negative long-term consequences.

Once substance abuse takes over, a person's ability to exert self-control can become seriously impaired. Brain imaging studies from drug-addicted individuals show physical changes in those areas of the brain that are critical to judgment, decision-making, learning and memory, as well as behavior control. Scientists believe that these changes alter the way the brain works, and may help explain the de-structive behaviors of addiction. The brain may no longer cooperate with what the mind wants. Addicts may think they are in control, but their addiction actually has come to control them.

This takes us to the next question. Is there still a way out for people caught up in the cycle of any substance or behavioral addiction? The answer is a definite yes! Addiction need not be a life sentence. Like other chronic diseases, addiction can be managed successfully if we *decide* to do something about it. Treatment enables people to counteract the addiction's powerful disruptive effects on brain and behavior in order to regain control over their lives. Research shows that combining behavioral therapy with treatment medications, where available, is the best way to ensure success for most patients.

As to the question of free will, I would conclude the following: We usually start addictions voluntarily and we have the ability to end them voluntarily. Self-control, also called self-mastery, is training in human freedom. The two alternative choices are clear: Either we govern our passions and find peace, or we let ourselves be dominated by them and become unhappy. We need to become masters of our feelings and emotions again—which is a life-long task.

Self-control is not inborn but must be taught and nurtured by ridding oneself of all slavery to unruly passions. But often it is hard to do all this on your own, alone. Fortunately, there are self-help AA-like organizations that aid people to overcome their addiction. Thanks to these, we can still be, or become again, masters over our own actions. Human behavior is more often than not a matter of lifestyle *choices* rather than the outcome of a set of genetic instructions. We are not just at the mercy of our genes. We encounter here a peculiar situation: Those who *believe* their addiction is a chronic and genetic brain issue are less likely to kick the habit.

14. No Gene for Rationality

In the previous chapter, I came to the conclusion that the mind is different from the brain, which goes contrary to the predominant mantra and prevailing paradigm of many life scientists, including geneticists. However, equating the mind to the brain is not a *conclusion* made after much research but an *assumption* before research even begins. It is a pre- or proto-scientific assumption. How come that the majority of scientists is not aware of this assumption?

Most scientists are strongly attached to the paradigm they were brought up with in their scientific education. That is how they learned their standards, by solving "standard" problems, performing "standard" experiments, and eventually by doing research under a supervisor already skilled within the paradigm. Because of this training, scientists usually do not question the paradigm's assumptions and are often even unable to articulate the precise nature of the paradigm in which they work... until they are forced to reflect on it further. That is what I am trying to stimulate here.

If my philosophical analysis is correct so far, then we have to face some important implications. First of all, we have to realize neuro-scientists are not mind-readers, neuro-surgeons are not mind-surgeons, and neuro-science is not mind-science. Simply put, thoughts are more than brain waves—in the same way as love is more than a chemical reaction. Medical professionals can read and interpret an electroencephalogram (EEG) or a magnetic resonance image (MRI), but looking at these does not show them any thoughts—perhaps memory "traces" of thoughts, but not the thoughts themselves. They just cannot "read" your mind.

The German philosopher Gottfried Leibniz once suggested to picture the brain so much enlarged that one could walk in it as if in a mill. Inside, we would only observe movements of several parts, but never anything like a thought. For this reason, he concluded thoughts must be different from physical and material movements and parts—they are not material but immaterial. Nowadays, the mechanical model of cogs and wheels which Leibniz used has been replaced by the chemical model of biochemical pathways, but the outcome is the same. If Leibniz is right—and I think I have given you already enough reasons to believe he is—that would explain why brain-scans never reveal thoughts; all they can pick up are "brain waves," but never thoughts, since those fail to show up on pictures and scans.

Even the fact that certain mental phenomena are associated with certain neural phenomena does not entail that these mental phenomena were *caused* by neural phenomena. Correlation does not equal causation. Take functional magnetic resonance imaging (fMRI)—an MRI procedure that measures brain activity by detecting associated changes in blood flow. When certain regions light up on an fMRI, that very fact does not explain whether this lit-up state is causing a certain mental state or just reflecting it. It could very well be, as we discussed earlier, that the brain is merely a material carrier for the mind's immaterial thoughts—so it would ultimately be the mind that makes those areas light up. Neuronal activities may be evidence of mental activities, but that does not imply they are identical to mental activities.

There is even more confirmation for my distinction between brain and mind. Although something like pain can be induced in a physical way, there is no evidence that experimental stimulation of specific neuronal areas would produce a specific mental state or a specific thought. What can be physically induced is something like emotions or feelings (animals have those too), or even memories that had been stored in the brain—including memories of thoughts once produced by the mind—because memories can be physically stored, similar to the way thoughts can be "stored" on paper. Thoughts, on the other hand, cannot be produced in a physical manner, let alone by electrodes. If the thought of "two times two" would physically produce the thought of "four," we could have skipped much time in school.

Where do these thoughts come from then? Yes, they come from the mind, but more specifically they originate from our capacity of *rationality*. What do I mean by "rationality"? Rationality gives us access to the world of truths and untruths—a world beyond our control; yet we can have an immaterial sense for what is true and what is false. Rationality is our capacity for abstract thinking and having reasons for our thoughts, thus giving us access to the "unseen" world of thoughts, laws, and truths. Rationality allows us to gain knowledge about the world through the power of abstract concepts and mental reasoning, thus giving us an immaterial sense for what is true and what is false. Weighing evidence and coming to a conclusion are rational activities par excellence.

Reasoning leads us from one idea to a related idea; it is a matter of pondering realities beyond that which we experience through our physical senses, thus allowing us to transcend the current situation with the mental power of abstract concepts and mental reasoning. As rational beings, endowed with the capacity of rationality, we are in pursuit of what is true (versus false). The word "Man" goes back to the Sanskrit word "Manu" meaning "thinking" or "wise"—for this is indeed the essence of our being. No wonder we are sometimes called "thinking animals." Our species name *Homo sapiens* was not a bad choice after all.

Philosophical giants such as Aristotle and Thomas Aquinas would put it this way: All we know about the world comes through our physical senses but this is then processed by the immaterial *intellect* that extracts from sensory experiences that which is intelligible. Well, it is the rationality of our intellect that makes the world intelligible and understandable; it gives us the power to comprehend the universe. Intelligibility is not a matter of intelligence but of intellect. If you ask me what it is in us that keeps searching for the truth, I would say certainly not the brain, but rather the mind's intellect that makes things "intelligible" for us.

It is the mind's rationality that gives us access to the laws of nature and the structure of this universe. Think of all those scientific explanations that are based on scientific laws, including genetic laws. They explain material things and events, and yet they use non-material laws of nature to do so. Right in the middle of our comfortable, spacious, temporal, transient, and piecemeal world of material things, something pops up that we call "laws of nature"—physical laws, Mendelian laws, mathematical laws, and even moral laws. Unlike all material things surrounding us, laws do not have any of the features that apply to the material world—that's right, none.

None, I said. A law such as the law of gravity is not located somewhere in space, not even in our minds, for that is just a mental "picture" of the law—but a law is everywhere. In a physical sense, laws are nowhere, and yet they are everywhere and apply to the entire universe; they are universal. Laws are also beyond time—timeless entities that cannot emerge nor perish in the history of the universe; we may discover them at a certain time, by they were already there before we discovered them. Neither are laws subject to change, for they will always remain true, even before we came to know them. And here is the most important difference: Not only are laws general, as we find the same law applied all over the universe, over and over again, but they also are necessary, which means that things in this universe cannot be different from what is expressed in those laws of nature. That makes them universal.

When scientists or engineers violate these laws, they get into real and actual trouble. A bridge that has been designed according to the right laws can stand firm, whereas another bridge collapses because its engineers erred in their calculations—perhaps they had the wrong laws in mind, or at least the wrong thoughts. The construction of a bridge would never depend on the right laws and the right thoughts, if those laws and thoughts were only creations of the human mind. It would not make sense to say that competent engineers have better mental habits than their inept colleagues. There must be more to it.

Now we should be ready to answer the question of why rationality cannot be in the genes. I am going to use a two-stage approach, one with an empirical basis and one with a philosophical basis. I hope this is not too much philosophy for the more scientific minds and not too much science for the more philosophical minds. I am taking a risk, though, for Albert Einstein was probably right when he said: "the man of science is a poor philosopher" (although Einstein was a great exception himself). I think it is important to bridge this gap. Given the realities of present-day research, most scientists are locked into specialties that are quite narrow in scope and so demanding of their time that they can barely keep up with what is happening on a larger scale.

Let me start with the former approach first. The *empirical* answer would be that rationality does not exist in the animal world, and therefore could not have come from there through genes in the course of evolution. What is it that makes me so sure, in spite of contrary claims of many evolutionists, that there is no rationality in the animal world?

Animals may be more or less intelligent, but they are not rational beings. Let me explain. Pets can even be smarter than their owners when they play a whole repertoire of tricks on their owner's emotions—but that is a matter of intelligence at best, not intellect. Animals do have the capacity to sense and remember things, but they lack understanding in the sense of asking questions, formulating concepts, framing propositions, and drawing conclusions. They show no signs of abstract reasoning or having reasons for their "thoughts" (if they have any); they do not think in terms of true and false; they do not think in terms of cause-and-effect with "if-and-only-if" statements. Instead, they are "moved" by motives, drives, instincts, emotions, stimuli, and training, but not by reasons or mental concepts. In other words, animals do not have an intellect endowed with the capacity for rationality—regardless of their intelligence.

Mental concepts are uniquely human. When we train a dog to associate a command such as "The boss!" with its real boss, then the dog has been conditioned to respond to such a command by looking for the real boss. It has become a signal, but signals are very different from symbols and mental concepts. Signals depend on the actual presence of the "real thing," due to associative conditioning (training). The dog has a physical image of its own boss, but it has no mental concept of what a "boss" is like.

Humans, on the other hand, can use the word "boss" also as a mental concept of "any boss." They often use that word to talk about what their own boss is like or should be like—preferably only when their physical boss is *not* around. Mental concepts transform "things" of the world into "objects" of knowledge, thus enabling humans to see with their "mental eyes" what no physical eyes could ever see before. Pets, however, do not chat about or meditate on what their boss is like—they are not pondering creatures. They can handle signals but not symbols.

Symbols refer, signals do not. It does not matter whether you train a dog with a command like "Here!" or a command like "Hector!" The dog reacts the same way, not realizing the latter command refers to himself. For animals, both commands work through association, but only humans know they are fundamentally different. Animals cannot make this distinction, so they treat everything in their surroundings as signals that call for a direct and definite response. Animals are "born positivists"—they take everything at face value. Humans, on the other hand, deal with things after making a "detour," through symbols and concepts; they extrapolate from what is seen to what is unseen; they assign various interpretations to the things they see. They move from the world of sensible singulars (things and events) to the world of immaterial universals (concepts and symbols). Whereas signals have their own intrinsic properties, symbols have a wider meaning and interpretation. That is the reason why *Homo sapiens* has also been called an *animal symbolicum*.

Animals do not have this capacity. When I warn a dog by pointing my finger to an approaching car, the dog just looks at my finger—and may even lick it—but it does not get what my finger refers to; it cannot make various interpretations. For animals, my pointing finger is not a symbol, but just what it is, a finger (although they can be trained to associate this with something else as a signal). My finger just cannot direct the animal's attention to something beyond itself, for that requires interpretation. A prey animal, for example, can only take a predator as a signal to flee or attack, but not as a symbol with various interpretations—for instance, as an animal in need, as an animal who is born to prey, as an animal brought up that way, as an addicted killer, as a preprogrammed killer, as a member of a larger conspiracy, as an inevitable part of life, or as a part of nature that needs to be preserved. Humans can come up with such different interpretations, but animals cannot, for they see only signals that call for an immediate response (true, we do too sometimes!).

Symbols are part of reasoning. Whereas reasoning is pondering realities beyond that which is experienced through the senses, animals, in contrast, seem to live their lives entirely in the present, without having any thoughts about the past or the future—perhaps memories, but not thoughts, symbols, or interpretations. If pets have a pedigree, it is thanks to their owners; if they have birthdays, wish lists, appointments, or schedules, it is because their owners create those; and if they have graves, those were dug by their owners as well. Cats or dogs have never come up with the thought of going to the pet store and buying their own food, let alone of starting their own pet store.

Since we are masters of anthropomorphism, we tend to think that animals have got to be like humans, even with regard to rationality. But what a disparity there is between them and us! Only humans are conscious of time; they can study the past, recognize the present, and anticipate the future; they even desire to transcend time, thinking about living forever. Only humans wonder "what caused or will cause what and why?" Only human beings have inquisitive minds asking questions such as "Where do we come from?" and "Why are we here?" Only humans have the capacity to be scholars and scientists; they can even study animals, whereas animals can only watch humans but never study them. Human beings are always in search of some kind of worldview or explanation of life—which certainly goes far beyond their need for food. In short, human beings are *questioning* beings; they are driven by rationality, which gives them the capacity to make rational decisions (without any guarantee, of course, that those decisions are always rational).

Rationality is not a matter of intelligence but of intellect. Whereas intelligence can be graded on an IQ scale, intellect cannot. One can have more or less intelligence but one cannot have more or less intellect. Intelligence only works with perception of sense-data. Consequently, animals may show various forms of intelligence in their behavior, because intelligence is a brain feature and as such an important tool in survival. We find spatial intelligence in pigeons and bats, social intelligence in wolves and monkeys, formal intelligence in apes and dolphins, practical intelligence in rats and ravens, to name just a few. Intelligence is a matter of processing sense-data—something even a robot can do by "cleverly" processing sounds, images, stimuli, signals, and the like.

Intellect is very different from this. Like intelligence, intellect also uses sense-data, but unlike intelligence, it changes perception into cognition by means of mental concepts and logical reasoning, which makes sensorial experiences *intelligible* for the human mind. A mental concept may be as simple as a "circle" or as complex as a "gene," but a concept definitely goes beyond what the senses provide. We do not "see" genes but have come to hypothesize and conceptualize them. We do not even see circles, for a "circle" is a highly abstract, idealized concept (with a radius and a diameter). Animals are certainly able to form general images, but they cannot form universal concepts, for the latter require human interpretation. (There is that difference between signals and symbols again.) Concepts go far beyond what the senses provide. Images, on the other hand, are inherently ambiguous because they can be interpreted in boundlessly different ways.

Mental concepts transform "things" of the world into "objects" of knowledge. It is rationality that makes the world intelligible and understandable; it gives us the power to comprehend the universe. Let us see what this entails. It is only thanks to our rationality that we take the world as an orderly and lawful and comprehensible entity. Rationality assumes order, stability, and predictability in this universe, for it is to the essence of rationality that there are truths built into us and into the world that reason can apprehend. These assumptions are not scientific output, but rather philosophical input; they do not come from science but they enable science.

Make no mistake, science could never prove that the world is orderly and lawful and comprehensible. Why not? If there were no order in the universe, it would make no sense to search for laws of nature in physics, chemistry, biology, genetics, and other disciplines. It is only due to the orderly design of the universe that we can explain and predict—which would be impossible in a world of disorder and irregularity. Since we may assume that like causes produce like effects, we are able to explain and predict. All of this is only possible thanks to the fact that we have an intellect capable of rationality.

Apparently, science in itself is inherently incomplete and a baseless enterprise, unless it receives a firm foundation from somewhere else. Let us discuss this first for the issue of *order* in this universe. Science can never scientifically prove there is order in the world, instead it must assume it. Order must come first for science to follow. Order is proto-scientific—an assumption of the intellect—and must come first before science can even get started. The tool of falsification, for instance, is necessarily based on this very assumption. Without order, there would not be any falsifying evidence. When we do find falsifying evidence, we do not take this as proof that the universe is *not* orderly, but rather as an indication that there is something wrong with the specific order we had conjectured up in our minds.

In other words, falsification is based on order and cannot be falsified by disorder; counter-evidence may falsify a specific theory, but not the principle of falsification itself. In utter amazement, Albert Einstein wrote in one of his letters, "But surely, a priori, one should expect the world to be chaotic, not to be grasped by thought in any way." Einstein often spoke of the "harmony of the universe"—actually one of the main pillars of science.

Something similar holds for the notions of design and functionality, especially so in the life sciences and the technical sciences. A heart and a pump, or an eye and a camera, can and do work because of the way our universe has been designed and outfitted with teleology. They are "successful" designs, given the way the cosmic design is—and that makes them functional and teleological. Without some kind of cosmic design in the background, they could not work at all (see Chapter 7). As a consequence, natural selection can only select those biological designs that are in accordance with the cosmic design. That is why functionality is a basic notion in the life sciences, making biologists look for the function of any biological feature—and if they cannot find one, they keep searching until they succeed. The assumption of functionality, of causes that have successful effects, is as basic to the life sciences as the assumption of causality is to all the natural sciences. Functionality is not an outcome of extensive research but a presupposition. All science can do is finding out how functional certain biological designs are—but the fact that there *is* functionality and design in the world of organisms is beyond its reach.

A third issue would be the enigma of this universe being intelligible. Where could this notion of *intelligibility* come from? Again, it certainly does not come from science itself. Scientists assume that, in principle, the world can be known and taken as intelligible—otherwise there would be no reason to pursue science. So intelligibility is definitely not the outcome of intense and extensive scientific research; it is not intra- but extra-scientific; it is a proto-scientific notion that must come first before science can even get started. It is so basic to science that it easily eludes scientists.

If you were told scientists had discovered that certain physical phenomena are *not* intelligible, you would, or at least should, tell them to keep searching and come up with a better hypothesis or theory—based on this fundamental philosophical assumption that says the universe is "fundamentally" intelligible and comprehensible. Rationality calls for it. Albert Einstein used to say that the most incomprehensible thing about the universe is that it is comprehensible, which is actually a mystery.

I would conclude from this that there is *empirical* evidence that rationality did not come from genes derived from the animal world, because animals do not have rationality. Of course, one could argue that rationality is just an illusion or that it is based on uniquely human genes not found in animals. That is where the second stage comes in—a *philosophical* argument as to why rationality cannot possible be in the genes. This philosophical argument would go along the following lines.

If rationality were the outcome of genes, subject to natural selection, then we would immediately run into philosophical and logical problems. We would bump into the following philosophical paradox: The theory of natural selection is the product of rationality—according to a hypothesis generated by Charles Darwin's rationality—and at the same time, rationality is supposedly the product of natural selection. Put differently, if Darwin's thoughts were the mere product of natural selection, so would be his science, and as a consequence, none of his thoughts—or ours, for that matter—could then be trusted.

Some might strongly object to what I am stating here. They would argue the following: Natural selection favors those who make correct interpretations, so those who believe what is true have a better chance of surviving than those who believe in something false; if natural selection did not operate based on truth, we could not exist. Therefore, we can trust our beliefs, for they went through the rigorous test of natural selection.

I do not think such objections are valid. My main reasoning goes along these lines. Natural selection does not operate on *truth* but on what is in accordance with *reality*; truths and truth claims are *intellectual* concepts of what reality is supposed to be like. There is no evidence that truth claims are determined by genes. Truth claims are immaterial assessments regarding the immaterial world of facts, not about the material world of things and events; facts are our *interpretations* of reality—of things, situations, and events around us. Facts are intellectual, rational entities. A fact is not an event but the description of an event, not a thought but the object of a thought, and not a statement but the content of a statement.

In other words, there are no facts without thoughts and statements. Natural selection may indeed favor those who make correct interpretations about the world around them, thus ensuring that those who believe what is true have a better chance of surviving. But not only do our interpretations change constantly, it is also very doubtful whether they are controlled by genes—and to the extent they are not, they cannot be subject to natural selection. Weighing evidence and coming to a conclusion are rational activities that cannot possibly be determined by genes; if they were, their outcome would be inevitable and predetermined. The history of science, for one thing, has repeatedly falsified this claim.

I cannot see how Newton's discovery of gravity, Mendel's discovery of genes, and Darwin's discovery of natural selection could have been catapulted by their genes, nor did the genes change of those who accepted these discoveries—they just deemed them to be true, on rational grounds, without any genetic interference. Facts are immaterial, mental creations that are beyond the reach of material genes. So I must come to the conclusion that our beliefs about facts are not anchored in the genes—this being one of the reasons why our beliefs can change. Rationality gives us the power to overrule any "innate, intuitive convictions" (if there are such things). And let's face it, most statements about facts—especially so in science—have no or hardly any survival value.

To claim that the theory of natural selection is the product of rationality, and at the same time, that rationality is supposedly the product of natural selection makes for a "boomerang theory" that undermines its own claims; it acts like quicksand—once we consider it to be true, it becomes false and thus lets us sink down. To use an analogy, a hand can make a drawing of a hand on paper but it cannot draw the very hand that does the drawing! We can only escape from this verdict if our rationality is something "more" than what natural selection has produced; to put it in an image again, the hand that draws a hand on paper must be more than the hand it is drawing—or likewise, the mind must be more than the brain it studies.

If our rationality were merely a brain issue—or more so an issue of genes—we would have no way of distinguishing between true and false. If the mental were the same as the neural, thoughts could never be true or false, as neural events simply happen, and that is that! So my conclusion is that we obviously do not need any sophisticated scientific equipment to debunk the quasi-scientific claim that there is a gene for rationality. Philosophy can tell us ahead of time that such a claim does not have much of a chance and may not be worth any further scientific efforts and financial funding. Genes just cannot produce truths and untruths. If I believe that my beliefs are the mere product of genes, then I have no reason to believe my belief is true—therefore, I have no reason to believe that my beliefs are the mere product of genes.

We have a similar problem if we would argue that the brain can be studied by the brain of a neuroscientist. The philosophical question would be: Could the brain ever study the brain all by itself? That would be like the magic of a projector projecting itself or a copy machine copying itself. Those who study the brain must be "more" than their own brains—in the same way as Watson and Crick must have been "more" than the DNA they discovered and studied. We just cannot pull ourselves up by our own bootstraps. To put it in more philosophical terms, the *knowing subject* must be "more" than the *known object*.

So I would say that only the mind as a knowing subject is able to study the brain as a known object, because it requires a mind to understand the brain, as it requires a subject to study any object. When studying the human brain as an object of science, a scientist needs the human mind as the subject of science—for without the human mind, with its intellect and rationality, there would be no science at all. The mind must be more than the brain it studies.

Even Charles Darwin vaguely realized this when he said in his Autobiography, "But then with me the horrid doubt always arises whether the convictions of man's mind, which has been developed from the mind of the lower animals, are of any value or at all trustworthy. Would anyone trust in the convictions of a monkey's mind, if there are any convictions in such a mind?" The theory of natural selection makes him wonder whether "the mind of man, which has, as I fully believe, been developed from a mind as low as that possessed by the lowest animal, [can] be trusted when it draws such grand conclusions."

Darwin would have been right with his skepticism had he said that our convictions cannot be trusted if they came from a *brain* that developed from a monkey's brain—but instead he erroneously used the word "mind," thinking they are the same. An animal has a brain, no mind, but a human being has more than a brain—a mind, that is. The brain may have developed from the animal kingdom, but it is hard to maintain that the mind is also a product of evolution. If it were, we definitely should be questioning the validity of our knowledge—which necessarily includes our scientific knowledge of the theory of evolution. Darwin did not seem to realize that the theory of natural selection must *assume* the human mind, but can neither create nor explain it. If the human mind were really the mere product of natural selection, so would be science, hence nothing we claim to know could then be trusted.

My conclusion is that philosophy can show us ahead of time, *a priori*, that the mind of a person such as Darwin, who discovered the theory of evolutionary theory, or of scientists such as Watson and Crick, who discovered the structure of DNA, must be more than that which they discovered. Otherwise their discoveries would not be discoveries, but very shaky and fragile claims, or not even that. They would be mere illusions concocted by a neural network under the direction of genes.

If rationality really were the product of evolution, favored by natural selection, our genes would only make us *believe*—surely a shaky belief, though—that our rational capabilities have an objective foundation, but in fact they would not and could not, for their only foundation would lie in our genes. So we would end up with a collective illusion foisted on us by our genes. We would have no *reason* to trust our own reasoning. That would be the end of anything we claim to be true—a thought that truly stops all thought.

I hope it is obvious by now that if natural selection were the origin of all there is in life, including the human mind, it would act as a boomerang that comes back to its maker, in a vicious circle, knocking out the truth claims of whoever launched it. How could we ever trust the outcome of mere natural selection when it comes to matters of truth? On its own, natural selection would be just a powerless and useless concept, for if one cannot trust the rationality of human beings, one is logically prevented from having confidence in one's own rational activities—with science being one of them.

We have here another case of a "boomerang theory"—once we consider it to be true, it becomes false. If I believe that my beliefs are determined by genes, then this very belief must also be determined by genes—which creates a paradox by violating the principle of non-contradiction. The trouble of claiming that rationality is in the genes is that it cuts off our reason for reasoning and for trusting our own rationality! To put it briefly, since there are no genes for truths and untruths, rationality cannot be in the genes. Truth and untruth is not just a matter of opinion—let alone of genes. People who do not care about what is true and what is false have *chosen* not to care; it is not their genes that made them choose that way. Instead of stating that genes determine what we call true or false, I would state the opposite: Rationality has the power to overrule what our genes dictate.

A gene cannot make something true—perhaps more or less effective, more or less successful, or whatever, but never true, let alone more or less true. What is true or false is not determined by genes—genes just cannot make anything true or false. Truth is not under the control of genes, but instead genes are under the control of what is true in our world (based on laws of nature). Therefore, we have the freedom to accept or reject what is true and what is false—as we see happening all the time—because truths and genes are of a different nature.

15. No Gene for Morality

Again I am going to follow two strategies to demonstrate that there is no gene for morality: the first one is to *empirically* demonstrate that there is no morality in the animal world, and the second one to prove *philosophically* that morality cannot be in the genes. But before I do so, I should explain first what morality is, and what it is not.

Let me stress first that morality is not another word for social behavior. They are very different notions. Whereas social behavior does have evolutionary roots in the animal world, morality does not. What is it then that makes morality so different from social behavior?

As moral beings, endowed with the capacity of *morality*, we are in pursuit of what is right (versus wrong). Morality gives us an "immaterial sense" for what is right and what is wrong. Not everything that is thinkable or possible or reasonable—in rational terms—is also permissible in moral terms. The term morality can be confusing, though. The (philosophical) study of morality is usually called ethics, and ethics has given us various interpretations of morality. I take morality to be about rights and obligations, about actions others owe us (our rights) and about actions we owe to others (our duties), both being part of the "common good." Morality gives us access to a world of duties and rights—a world beyond our control, although we do have an immaterial sense of what is morally right and what is morally wrong. It adds a very different dimension to social behavior.

Morality is more than a series of do's and don'ts; it is more than the capacity of making moral decisions. Morality most specifically guides us as to which actions we owe to others (human obligations) and which actions others owe us (human rights). Morality does not come with a special race, nation, party, or church—it is a common property that belongs to us all. Duties and rights have a natural reciprocity: The duty of self-preservation is also the right of self-preservation; the duty to seek the truth is also the right to seek it; the duty to work for justice is also the right to pursue it. In other words, no duties no rights, and no rights no duties. No one has the duty to have children, so no one has the right to have children.

So morality is not about what the world *is* like, but about what the world *ought* to be like; it is not a matter of description but prescription. It is not a description of what social behavior *is* like, but a prescription of what social behavior *should* be like. "Racial equality," for instance, is not a descriptive but prescriptive term; races are not equal in biological characteristics but they do have the same dignity and rights. Morality tells us what *ought* to be done in life—no matter what, whether we like it or not, whether we feel it or not, whether others enforce it or not. It tells us what ought to be done—by us, as a duty, and towards us, as a right—otherwise a moral mistake would be made. It is essential to all of us that we can discern if human beings are doing the right thing or not. Whereas our movements are subject to physical constraints, our actions are subject to moral ones.

There are some strong resemblances between morality and rationality. They both are universal (applicable to everyone everywhere), absolute (without exceptions), eternal (even if we do not know the law behind it yet), and objective (a given, independent of us and of any human authority). They are objective, universal, eternal, and absolute standards—no matter whether we are talking rationality, in terms of true and false, or morality, in terms of right and wrong. Here are more details on those similarities.

1. Rational rules as well as moral obligations are evident. It is plainly evident that there is order in this world and that like causes produce like effects—there is just no hard proof for it (see Chapter 14). It is equally evident that it is wrong to kill other human beings—there is nothing we can come up with in support of it.

2. Rationality is in search of universal laws and objective truths in this

universe as these tell us the way it is in this world—no matter what, whether we like it or not, whether we feel it or not. "Truths are true," even when we do not know yet they are true. Truths are not dependent on our knowledge and are not created by our knowledge. In a similar way, morality is in search of universal obligations and objective values in this world; these tell us what we ought to do—no matter what, whether we like it or not, whether we feel it or not. "Rights are right," even when we do not know yet they are right. Just as there are physical laws that may work but can still be wrong, there are moral laws that may work but can still be wrong.

3. Natural laws and moral laws are both absolute. Scientists are basically absolutists: They are ultimately in search of absolute, objective, universal laws of nature, but they realize they may not have reached that point yet. In morality, we should strive for something similar. Just as we may be oblivious to laws of nature that we do not know yet, we may violate moral laws we are not aware of yet. Relativists in morality, on the other hand, defy themselves when they make the absolute statement that everything is relative. Deciding on what is true and what is right is not a matter of opinion.

4. We cannot even establish rationality by showing how irrational it would be to reject rationality, since that would presume already some sense of rationality. Neither can we establish morality by pointing out how immoral it would be to reject morality, for that would already require a basic sense of morality. As human beings we have a sense of rationality and a sense of morality—inborn if you will, but not genetic.

5. Our genes do not determine how we choose between true and false, or between right and wrong. It seems to be quite the opposite: Rationality and morality have the power to overrule what our genes dictate. In fact, they both persistently attempt to distance us from what we are or would be if our genes were in full control.

6. If you have ever been on jury duty, you know how much jurors de-

pend on their human faculties of rationality and morality. Unlike animals, we are able to make rational and moral decisions—which fact also entails the possibility of irrational and immoral decisions, of course. Take rationality and morality away from us, and we are back where we came from—animals. Only human beings can curb their animal drives and instincts with rationality and morality.

Let us focus exclusively on morality again. There is something peculiar about morality: We cannot define moral (prescriptive) notions in non-moral (descriptive) terms; the fact that something *is* a certain way does not necessarily entail that it *ought* to be that way. Or to put it more concisely, description does not automatically lead to prescription. Here are some examples. The fact that diseases are "natural" in a biological sense does not entail they are "good" in a moral sense too— that's why we ought to fight them and cure them. The fact that there is biological development does not mean there is also a development in human dignity. The fact that the "survival of the fittest" may be natural does not mean that we should enforce it morally. Morality actually demands that we give the same care to the weakest in society as we give to the strongest. And the fact that some people are richer than others or more intelligent than others does not mean that we ought to value them differently in a moral sense. They may have more power than others, but they should not have more rights or fewer duties than others. The fact that human beings *are* different does not mean they *ought* to be treated in a different way. Hence, a moral property such as being good or right cannot be reduced to a natural property such as being natural, functional, genetic, more evolved, more developed, better for the majority, or whatever.

Let me explain this point further with an example used by our former president Abraham Lincoln, of all people. It seems to me that Abe put things better than I ever could when he was talking about slavery. In his own, rather technical words,

> "If A. can prove, however conclusively, that he may, of right, enslave B.—why may not B. snatch the same argument, and prove equally, that he may enslave A?—You say A. is white, and B. is black. It is color, then; the lighter, having the right to enslave the darker? Take care. By this rule, you are to be slave to the first man you meet, with a fairer skin than your own. You mean the whites are intellectually the superiors of the blacks; and, therefore have the right to enslave them? Take care again. By this rule, you are to be slave to the first man you meet, with an intellect superior to your own."

President Lincoln's point is clear: All the answers some people might come up with to defend their moral claims use relative criteria that are morally irrelevant besides, such as a darker skin color or a lower intelligence. Because those criteria are relative, someone with a lighter skin or higher intelligence would then have the "moral right" to enslave you. And the same holds for the moral value of human life. This value cannot be based on biological standards such as age, viability, vitality, or brain volume, since those are per definition morally irrelevant, and relative besides.

In contrast, moral values are absolute ends-in-themselves—not disposable means-to-other-ends. In morality, there is no "Thou shall not…, unless…" Whereas our bodily movements are subject to physical constraints, our social actions are subject to moral ones. As I said earlier, description does not lead to prescription; however, prescription must be rooted in description. We cannot be morally obliged to something that is physically impossible. The only authority that can obligate us is something—or rather someone—infinitely superior to us; no one else has the right to demand our absolute obedience in matters of human dignity, human freedom, and the like. Moral rights and duties are absolute, objective standards of human behavior—they are nonnegotiable. We are fully responsible for the moral choices we make in life.

Indeed, we do have a choice when it comes to morality, but that does not mean we can just pick whatever we want. We cannot just vote to decide whether we are anti-slavery and anti-abortion, or not. Abraham Lincoln put it well when he challenged the Nebraska bill of 1820 that would let residents vote to decide if slavery would be legal in their territory: "God did not place good and evil before man, telling him to make his choice." There is no "pro-choice" in morality.

You might object that moral values and laws are far from universal and absolute. As a matter of fact, many people claim that moral values have been subject to change during the course of human history—determined by a majority vote, so to speak. However, I would point out that we should not confuse moral *values* with moral *evaluations*. Moral evaluations are our personal feelings or discernments regarding moral values. Yet, some people think that, in making evaluations, we create values in accordance with these evaluations. So when evaluations change, the moral values and laws are said to change as well. If that were true, our moral values would indeed be subject to various cultural and historical fluctuations; morality would just be a matter of emotions, personal preferences, cultural trends, political powers, and majority votes.

In response, I would emphasize that evaluations are merely a reflection of the way we discern moral values and react to them—but do not mistake our evaluations for values. Whereas moral evaluations may be volatile and fluctuating, moral values and laws are eternal, universal, objective, and absolute. Think of the following comparison: Our current understanding of physical or biological laws constantly needs revision each time we reach a better understanding of those laws in the way they really are. In the meantime, we assume there are absolute and universal laws of nature, although we may not yet have fully captured them in our current understanding and our existing evaluations. Something similar holds for moral laws. As C.S. Lewis once put it, "The human mind has no more power of inventing a new value than of imagining a new primary color."

Let me illustrate this point a little further. A few centuries ago, slavery was not evaluated as morally wrong, but nowadays it is by most people. Did our moral values change? No, they did not; but our evaluations certainly did. Only a few people in the past were able to discern the objective, intrinsic, and universal value of personal freedom and human rights (versus slavery), whereas most of their contemporaries were blind for this value. That is the reason why Martin Luther King Jr. called an unjust (legal) law "a code that is out of harmony with the moral law." The law of the land is not always a reflection of the moral law. Moral laws are intrinsically right, even when we do not see yet that they are. Anyone who does not see their evidence is morally blind. Just like there are color-blind people, there are also value-blind people.

In spite of all of the above, some might object that morality could still be rooted in our genes. It is in fact a very common claim (later on in this chapter, I will discuss the more specific claim that even altruism must be anchored "in the genes" according to some socio-biologists). They say that we are "hard-wired" to be moral and make moral judgments. There even seems to be empirical evidence that 6 to 10-months old overwhelmingly prefer "good guys" over "bad guys." So the question is: Are we really hard-wired to identify that some things are good and some are not?

Well, here is my first strategy to debunk the claim that human morality stems from the animal world and was handed down to us through genes and DNA. My competing, empirical claim is that there is no morality in the animal world. Animals do not have any moral values, so they do not have to control their drives, lusts, and emotions. They just follow whatever "pops up" in their brains—and no one has the right to morally blame them. The relationship between predator and prey, for instance, has nothing to do with morality; if predators really had a conscience guided by morality, their lives would be pretty harsh. Dogs may act as if they are "caring," but they just follow their instinct, not some moral code; dogs happen to have such an instinct, whereas cats lack it, since it is not in their genes.

As a consequence, animals never do awful things out of meanness or cruelty, for the simple reason that they have no morality—and thus no cruelty or meanness. But humans definitely do have the capacity of performing real atrocities. Even kids who bully others should be held accountable and disciplined, regardless of their age. On the other hand, if animals do seem to do awful things, it is only because we as human beings consider their actions "awful" according to our standards of morality. Yet, we will never arrange court sessions for grizzly bears that maul hikers, because we know bears are not morally responsible for their actions.

Whereas animals do have social behavior, they have no moral values—which means no duties, no responsibilities, and consequently no rights. If animals had rights, their fellow animals would also need to respect those. We, as masters of anthropomorphism, may think animals are bound to have morality, but they are not. On the other hand, since we, as human beings, do have morality, we need to treat animals, God's other creatures, humanely and responsibly—not because animals deserve it, but because humans owe it to their Maker and to themselves, being stewards of what their Maker created. For this reason, scientific experimentation on live animals (vivisection) has ethical restrictions; in many jurisdictions, use of anesthesia is legally mandated for any surgery likely to cause pain to any vertebrate. Although animals are not human, *we* certainly are, so we ought to treat them humanely.

Apparently, human beings are moral beings, quite unlike their fellow animals. Let me repeat it again: Our social behavior can have evolutionary roots in the animal world, but morality cannot. Yet, morality is at the core of our being. If we know already at a very young age that there is right and wrong, good and evil, that does not necessarily mean we are "hard-wired" to be moral, but that morality is at the core of being human and is distinctively human. In that particular sense, morality is "inborn"—part of our constitution, without being hard-wired in our genes and brains.

Now it is time for my second strategy—to prove philosophically that morality cannot be in the genes. If morality were in the genes, why would we need an articulated moral rule to reinforce what "by nature" we would or would not desire to do anyway? Instead I would argue the opposite: Morality has the power to overrule what our genes dictate. The verdict as to what is morally right or wrong is not regulated by genes but by a moral code. The rules as to what is right and what is wrong do not and cannot come from genes. If moral behavior were genetic, we would not need a moral code additionally.

Reality tells us that far too many people are willing to break a moral rule when they can get away with it! Too many parents ignore their so-called "natural" responsibility. Too many spouses violate the sixth commandment, "You shall not commit adultery." Too many folks also violate the fifth commandment, "You shall not kill." When it comes to moral laws, everyone knows about them and yet everyone breaks them. Unlike the laws of nature, moral laws can in fact be ignored (try to do that with the law of gravity!). Mothers who abandon their babies are very unusual in the animal world, because of genetic constraints, but in human societies they are not so unusual, because maternal responsibility is a moral law that can be ignored.

Do those who go against moral laws or ignore them really go against their genes? Some biologists and philosophers think so. They suggest that evolutionary biology can explain how humanity acquired its morality. Their 'magic wand' is the theory of natural selection again. Some use the example of incest. There is an almost universal human taboo on incest—phrased as a moral law it says "intercourse with very close relatives is wrong and hence forbidden." Well, these biologists would point out that inbreeding between close relatives tends to bring out recessive lethal traits and other afflictions that lessen the offspring's reproductive success. Hence they argue that natural selection has been promoting a genetic basis for behavioral avoidance of intercourse with close relatives; and that is where they believe this moral value stems from. In their view, I assume, those who do commit incest must have a different allele somewhere.

Another example would be the moral laws given in the fifth and sixth commandments of the Decalogue, "You shall not kill" and "You shall not commit adultery." Some biologists have made the case that humans have become monogamous "by nature," since that would give the offspring a better protection, and that is why natural selection must have promoted monogamy—giving the sixth commandment a biological basis. Others have made the claim that killing any members of the same species would undermine the persistence of the species—hence the prohibition was selected for by the process of natural selection. They even consider the moral value of paternal care for children to be a product of natural selection, since fathers who do not feel an "instinctive" responsibility towards their underage children would reduce their offspring's reproductive success. What all these cases have in common is that morality is supposedly based on genes promoted by natural selection. They tell you there is nothing moral about moral values since it is all in the genes!

This way of thinking becomes even harder to defend when it comes to murder, especially mass murder. It is very unusual for animals to kill members of their own species—which is probably a genetic feature—but humans do kill their own. Just think of wars, weapons of mass destruction, concentration camps, fire squadrons, gang murders, geno-

cide, terrorism, and the list goes on and on. Did humans lose that protective gene or allele that their animal ancestors had? Or could mass
murder be the result of a positive selection pressure? Either case is
hard to defend. I would say instead that humans need a fifth commandment to stop them from committing atrocities. Moral laws such
as "You shall not kill" or "You shall not commit adultery" could never
make it through natural selection, for their offenders—the killers, the
promiscuous, and the rapists—would do much better in reproduction
than their victims or the ones who go by a moral code. Moral laws do
not have any survival value and therefore cannot be promoted by natural selection.

In contrast, I would argue that moral laws tell us to do what our genes
do *not* make us do by nature. Reality tells us that far too many people
are willing to break a moral rule when they can get away with it! Morality is about something that is outside the scope of biology, actually
beyond the reach of science. Biology is blind to moral values, so it
cannot possibly discern anything that is on its "blind spot." Science
cannot control morality, but it is actually the other way around—
morality should control science instead. Nazi-doctors such as Joseph
Mengele show us what happens when morality does not control rationality.

Nonetheless, biologists and geneticists keep trying to control morality
by reducing something moral to something biological. Probably the
best known and most controversial example comes from the biological
study of altruism. Let us find out how strong their case is. (But I must
note first that altruism is not a moral issue in the strict sense, because
it is not something that ought to be done; it is neither a right nor a
duty in a moral sense.)

You might think altruism would be safe against the attacks of evolutionists because altruism goes against the very idea of natural selec-

tion. Altruism is considered to be unselfish behavior—a sacrifice of personal comfort for the *benefit* of others—whereas natural selection is based on the principle of increasing one's own reproductive success at the *expense* of others. They cannot go together, can they?

At first sight, altruism does seem to be a problem in relation to natural selection. But socio-biologists such as E. O. Wilson did find a biological explanation. The pressing question in socio-biology is the following: How can altruistic behavior in the animal world still be advantageous to its agent? Well, socio-biologists did come up with some answers which I will not discuss in detail. All I want to gain is a better understanding of the logical and philosophical confusions we could be in for in this discussion. I will try to do so by introducing some clear distinctions.

First of all, there is a biological version of altruism—which I call *bio-altruism*. In the animal world, we do indeed find many examples of animals helping one another; just think of sterile worker bees "unselfishly" helping the queen raise her own progeny. Socio-biologists would explain bio-altruism as a form of helping one's close relatives, because these carry DNA very similar to one's own. Since natural selection is a matter of balancing the benefits against the costs, bio-altruism is a way of promoting one's "own" DNA by diminishing one's own offspring (a cost) but increasing the offspring of one's relatives (a benefit). And that is exactly what bees accomplish. Because these Hymenoptera have a very peculiar sexual system, females are more closely related to sisters (sharing 75% of their genetic material) than to daughters (sharing only 50%). So sterile females actually increase their own reproductive success by 25% in an "indirect" way—by helping rear their queen-sister's progeny (75%) rather than their own (50%).

This biological approach has many more applications than you might think. As the geneticist J. B. S. Haldane once said, "I will lay down my life for two brothers or eight cousins." His calculation of costs and benefits was based on the fact that in a regular sexual system, siblings share ½ of their genetic material, whereas cousins only share ⅛. By helping close relatives, one is somehow promoting dispersal of one's own genetic material, but in an indirect way through relatives For instance, through two brothers (2 x ½=1), or 8 cousins (8 x ⅛=1). This might be called altruism, but I would rather dub it as *bio*-altruism.

Secondly, there is also a sociological version of altruism—which I propose to call *socio*-altruism. It is based on the principle of helping those who return the help; divided you may fall but united you may conquer—something like "I give so you give." This happens, for example, when two or more organisms, such as chimpanzees, band together and thus, in helping others, they help themselves. This requires some form of intelligence (not intellect, though; see Chapter 14). Since there is social behavior in the animal world, there may also be socio-altruism. Socio-biologists have studied this phenomenon with what they call "game theory," and they have been rather successful doing so. I would say this is a success for socio-biology. And let us leave it at that.

Socio-biology may have solved the problems of bio-altruism and socio-altruism, but my question is: Has it also solved the problem of altruism in ethics—which I prefer to call *moral*-altruism? I do not believe so. Moral-altruism is a moral concept that is altogether different from bio-altruism and socio-altruism. Let me demonstrate their differences.

1.   Bio-altruism is behavior with the *effect* that one's own offspring is diminished but compensated for by helping relatives. So it may be functional and advantageous at times, and therefore subject to natural selection. It happens "by nature," and follows laws of nature.

2.   Socio-altruism is behavior with the *motive* of helping others, but limited to those who return the help. In helping others, one helps oneself according to the Roman motto *Do ut des*. Since it may enhance survival value, it may be subject to natural selection. It happens "by

nature." The development of cooperative behavior amongst animals makes perfect evolutionary sense.

3.   Moral-altruism, on the other hand, is behavior for the sake of the *value* of serving others, without expecting any advantage. This is what we really and rightly call "unselfish altruism." It is very likely that moral-altruism is partly based on the biological feature of socio-altruism and cooperation, but it also goes far beyond it. The notion of charity, for instance, is completely about giving for giving's sake. Or take donating blood to strangers: It does not help relatives, nor does it help banding gang members; it is not bio-altruism nor socio-altruism; and it is not subject to natural selection. Moral-altruism does not happen "by nature," but is guided by a moral code. It is in constant battle with our urge to survive, care for ourselves, and provide for our offspring. In short, moral-altruism is no good friend to survival; it amounts very often to "genetic suicide."

Let me use the following simple example to clarify the difference between a biological, social, and moral approach: What is wrong with bullying (or any other kind of violent behavior, such as rape)? As a biological feature, it may be very advantageous to have such a gene or allele. As a social strategy, it may be a very effective strategy of banding together against someone else. But as a moral issue, it is plainly wrong.

As we discussed earlier, biology, or one of its branches socio-biology, cannot have a monopolistic claim on human behavior, because biology will never be able to tell us a comprehensive, all-inclusive, story about human life and human behavior. Biologists approach everything only from a *biological* perspective—the rest of the story comes from other fields such as physics, psychology, economy, philosophy, religion—and ethics, of course. Since moral values add their own, new dimension and perspective to our world, there is not much hope for those numerous efforts of fully converting moral behavior into a non-moral phenomenon.

Yet, there are scientists who keep insisting that morality is an evolutionary product. They mention how morality is located in, and even regulated by, anatomical structures that we received from our animal ancestors—such as the prefrontal cortex, the orbital frontal cortex, the amygdala, and the list goes on and on. I don't think their claims are very impressive. It is like saying that a news report on the radio comes from an antenna, transistors, and a loud speaker. These do indeed help transmit the news but are certainly not the origin of the news. They can undergo improvement, but they do not determine the content of the news they broadcast.

Apparently, genetic determinism just does not and cannot apply to morality. Instead, you could even make the case that genetic determinism is one giant alibi for human responsibility. Once we declare ourselves no longer responsible for our moral decisions, we think we are off the moral hook. If we wish to be innocent, we must find a way to make the claim that we cannot be held morally responsible. And some have found a way! All you have to do is "geneticize" or "medicalize" bad moral behavior: The victimizer is no longer a person but a disease or pathology caused by genes—a "disease" supposedly beyond our moral control. Once we have made ourselves victims, we feel released from moral responsibility, since victims are, by definition, not responsible for their conditions, but can point instead to something else as the culprit—genes, hormones, diseases, syndromes, and pathologies. "My genes made me do this!" Morality, on the other hand, invites us to be more *self*-determined and less *pre*-determined. It tells us not to make excuses but to make commitments.

So we must come to the conclusion that social behavior may very well have a genetic component, but moral behavior cannot possibly be under genetic control. A gene cannot make something obligated— perhaps more or less effective, more or less successful, or whatever, but never more or less obligated, let alone morally right or wrong. What is right or wrong in a moral sense is not determined by genes— genes cannot make anything right or wrong. Genes may make us act a certain way, but whether such an act is morally right or wrong is a different issue—a moral issue, not a genetic one. Genes can make things either possible or impossible, but never right or wrong—the latter of which requires a moral system of laws, values, and evaluations. To use an analogy, genes may help create good or bad volleyball players, but not the rules of the game.

We do not have a moral "nature" if understood in terms of physiology and genetics. People who act immorally do not go against their "nature." And people who do not care about what is right and what is wrong have *chosen* not to care; it is not their genes that make them choose that way. Therefore, we have the freedom to accept or reject what is right and what is wrong—as we see happening all the time— because right and wrong are not determined by genes. Morality gives us the power to overrule any "innate" drives and emotions if we deem them immoral, but that choice is up to us.

16. No Gene for Religion

The discussion about a gene for religion has become very animated due to a 2005 book of the human geneticist Dean Hamer that he gave the provocative title *The God Gene: How Faith Is Hardwired into Our Genes*. The god-gene hypothesis was basically invented by Hamer, who now claims he has in fact discovered a gene that he decided to call the "god gene." To be more precise, he is talking about a gene for spirituality, which he deceptively dubbed as a "god gene."

Hamer theorized that if our sense of spirituality has a genetic basis, then those who rank higher in spirituality should share some genetic link that those who ranked lower do not. What did he mean by "spirituality"? He measured it by using a "self-transcendence" scale developed by the psychologist Robert Cloninger, in order to quantify how "spiritual" someone is, assuming that spirituality can be quantified by psychometric measurements. What impressed Hamer is that the self-transcendence measure had been shown to be heritable by classical twin studies. As an aside, those twin studies actually found that specific religious beliefs do *not* have a genetic basis but are transmitted by non-genetic means, such as culture, tradition, and imitation—so they could be inherited without being genetic.

What about the term "self-transcendence"? It is a word used by psychologists to describe spiritual feelings that are independent of what they call "traditional religion." Hence it is not based on belief in God, frequency of prayer, or any other conventional religious practice. Self-transcendent people tend to see everything, including themselves, as part of one great totality. They have a strong sense of "oneness" with people, places, and things. Self-transcendent individuals are also considered "mystical." They are fascinated with things that cannot be explained by science. They are creative but may also be prone to psychosis. In short, they are "spiritual," if you redefine this in a certain way.

In order to identify some of the specific genes involved in self-transcendence, Hamer analyzed DNA and personality score data from over thousand individuals. He asked them to fill out a detailed questionnaire—a standard test called a "Temperament and Character Inventory"—including a section that asked them to rate their feelings of "absentmindedness, connectedness with nature, belief in extrasensory perception, and other traits." He assumed that the answers would provide a measure of the subjects' affinity for what he called spirituality.

Then he went poking around in their genes to see if he could find the DNA responsible for their differences. With over 35,000 genes and 3.2 billion chemical bases in the human genome, he limited his search for the "spiritual gene" to nine genes known to produce monoamines (brain chemicals that regulate mood and motor control) and then identified one particular gene, VMAT2, as showing a significant correlation with affinity for spirituality. VMAT2 is a gene that codes for a vesicular monoamine transporter that plays a key role in regulating the levels of the brain chemicals serotonin, dopamine, and norepinephrine. These monoamine transmitters are in turn postulated to play an important role in regulating the brain activities associated with mystic and spiritual experiences.

When he analyzed this gene further, he discovered that those with the nucleic acid cytosine in one particular spot on the gene ranked high in spirituality, whereas those with the nucleic acid adenine in the same spot ranked lower. So he concluded that a single change in a single base in the middle of the "god gene"— at position 33050 of the human genome map, to be precise—seemed directly related to the ability to feel self-transcendence. He even gave an explanation as to why the "spiritual" allele for this gene would give its carrier a selective advantage: Spiritual individuals are believed to be favored by natural selection because they are provided with an innate sense of optimism, which produces positive effects at either a physical or psychological level.

What are we to make of all of this? Let me mention first that Hamer rushed into print with his book without any peer review and without publishing his results in a credible and reputable scientific journal—plus, which is even more serious, his findings have not been replicated. All of this probably gives us ample reason to not take his work on face value—or to put it nicely, it definitely deserves further scrutiny.

I could apply my previously used philosophical argument again: If I believe that my (religious or spiritual) beliefs come from genes, then my belief cannot be true, hence I do not have to believe that my beliefs come from genes. But in this case, I prefer to use a different strategy, to be done in three steps. First, I want to critically analyze Hamer's genetic claims. Second, I would like to show that religion—or more in general, a belief in God—cannot be reduced to something else than religion, for then we are no longer talking religion. Third, I want to demonstrate that believing in God, as part of religion, is more than a mere belief or spiritual feeling without any basis in reality.

Step one—a critical analysis of Hamer's claims. First of all, gene VMAT2 appears to be involved in the transport of monoamine neurotransmitters across the synapses of the brain. So it is basically a "pump" responsible for packaging a neurotransmitter for export during brain activity. In that specific sense, it is an important gene, and its product may even be active when you have "religious experiences," but that does not make this gene a "god gene." It is rather obvious that concepts of God do reside in your brain—they certainly do not reside in your toe—but that does not tell us where they ultimately come from. As an aside, I would even argue they do not come from your brain but from your mind. As I said earlier, the mind could very well *use* the brain, including its neurotransmitter "pumps."

Second, there are serious reasons why twin studies can be very deceiving, as we discussed earlier (Chapter 4). What twin studies about the link between genes and religion have shown us is that, in a particular cultural setting, a particular combination of genes and alleles can affect one's religious tendencies. Putting these genes in a different genetic or cultural environment could have a very different effect, though. When it comes to religion, the cultural environment is particularly important, because religion is also a trait of our culture and personality. So if we cannot even link individual genes to personalities, how can we possibly link them to religion? There are too many intervening steps involved—such as other genes, environmental effects on gene expression, cultural factors, and personal factors—to make such a simplistic link. Earlier we said that when something is partly hereditary, that does not mean it is also genetic. Even if we admit that genes do have an effect on how religious you are, upbringing still might have a big impact on the "brand" of religion you take on.

Third, Hamer has proved himself to be an expert in inventing genes— once it was a "gay gene" (see Chapter 4), now a "god gene." This reeks of some kind of political agenda. Unfortunately for him, the field of behavioral genetics is littered with failed links between particular genes and behavioral traits. We have been bombarded with new genes: a gene for alcohol addiction, a gene for homosexuality, a gene for schizophrenia, a gene for altruism, and now even a gene for religion—the list could go on and on.

These hypothetical genes are supposed to control traits with very complicated and variable patterns of behavior. They make for very complicated similarities that come in many varieties. Most of them were once claimed, and then had to be retracted; repeatedly they turned out to be inventions that did not lead to discoveries. Curiously enough, Hamer's "god gene" accounted for only 1% of the variance in the test scores of his subjects, prompting Francis Collins, former head of the *Human Genome Project*, to characterize Hamer's thesis as "wildly overstated." As someone else said, given the fate of Hamer's so-called gay gene, it is strange to see him so impatient to trumpet the discovery of his "god gene."

Fourth, even if we appear to be genetically hard-wired for religion, what could this mean? Clearly, we are not hard-wired for a particular religion; there are more than 7,000 identified varieties. "Born a Catholic" does not mean always a Catholic. Plenty are the cases of people who, in the course of their lives, decided to become atheists or decided the opposite by leaving atheism behind. Even if some part of spirituality is wired in the brain, the forms and practices of religion are still cultural and can be passed from one person to another by learning or imitation and can be changed by further experiences.

Fifth, if a certain allele were indeed the cause of being spiritual, we should expect that the number of people possessing this spiritual allele should at least be proportionate to those who consider themselves spiritual. If the number is disproportionate, the hypothesis of a "god gene" becomes suspect. In addition, all those possessing the "right" allele should have spiritual experiences; otherwise the presence of cytosine cannot be the cause of spirituality. Hamer failed to test for any of these ramifications.

Sixth, all of this genetic talk has actually nothing to do with God. So the label "god gene" is very deceiving, to say the least—which Hamer did acknowledge himself, though. In general, I would say that genetics can tell us a lot about human beings, but it has little or nothing to say about God. Perhaps genetics can tell us something about mystical experiences, but the idea that people believe in God because of mystical experiences is foolish. One need not feel anything, let alone have a mystical experience, to believe in the existence of God. Arguably, most individuals who believe in God have never experienced God in a mystical way. Quite a few believe in God for purely intellectual reasons. Others simply have an intuitive awareness of God's existence.

Seventh, the same holds for a "feeling of transcendence." Such a feeling is not necessarily a religious experience, and if Hamer is right, it is merely a biological one. The monoamines involved in the feeling of self-transcendence are the same monoamines that are scrambled by ecstasy, LSD, and other mind-altering drugs. If the feeling of transcendence is a biological experience rather than a religious experience, then studies performed on that experience only tell us something about biology, not religion. They are more related to emotionality than to rationality. The question of God's existence is not a biological issue, but rather a philosophical one.

Step two—reducing religion to something else than religion makes it no longer "religion." Hamer is an expert in mixing up very heterogeneous concepts. He speaks of "self-transcendence," "spirituality," "religion," "mysticism," "faith," and "God" interchangeably, but these terms are in fact quite different. Being spiritual does not coincide with being religious, for instance—let alone with believing in God. He actually redefines religion in spiritual terms as being part of one great totality.

To be fair, even Hamer himself admitted that those who possess cytosine, who have experienced self-transcendence and mysticism, do not necessarily believe in God. But if Hamer's gene is really a "god gene," one would expect that the approximately 90% of Americans who say they believe in God should have the allele with cytosine on their VMAT2 gene. Apparently, it is not religious faith or a belief in God that is hardwired into our genes but at best a single personality trait as measured by a standard psychological questionnaire. Well, those so-called self-transcendent persons may or may not believe in God.

This prompts the question of why Hamer wants to reduce religion and faith in God to something else, to something like spiritual experiences. The answer can be found in one of his main presuppositions—that biology can fully explain everything in life, including religious faith and experience. This presupposition is based on the view that all reality can be reduced to a scientific explanation and is only a matter of material entities. In this particular case, all human behavior can and must be fully explained in terms of biological functioning.

It should not surprise us that reducing religion to something else than religion has serious consequences. If there is only material stuff in this universe, there is no longer room for spiritual reality—perhaps for spiritual experiences but not for a reality behind and beyond those experiences. Hence, the "spiritual allele" of the "god gene" is not spiritual at all and can certainly not cause a religious experience. It may cause a *natural* experience of self-transcendence that some have unwittingly interpreted as an encounter with the *divine*. Ironically, Hammer seems to leave room for something "spiritual," but then he explains it away as merely something material. No wonder, he keeps hammering on the fact that his book is about whether "god genes" exist, not about whether there is a God. Does that take him off the hook?

The question of whether there is a God is not dependent on our experience of Him, and our experience of Him is not limited to a sense of self-transcendence. Although Hamer does not say whether there is a God or not, his philosophical presuppositions prevent him from even allowing that God might exist. While Hamer's work is not necessarily atheistic, it is motivated by the atheistic philosophy of materialism, which says that *matter* is all there is. This almost automatically leads to the idea that it is the need for God that created the idea of God. It was not the first, and by no means will it be the last attempt to relegate religious faith to a by-product of evolution, with no basis in reality.

Hamer stands in a long tradition of people who declare the God of religion a product of the brain—controlled by biological or psychological mechanisms. When religion tells us that we were made in God's image, supporters of this tradition would point out that God was actually made in our own image, made according to our own needs. Hence, religion ends up being an illusion—as Sigmund Freud would say—or a delusion—as Richard Dawkins describes it.

Hamer seems to fit perfectly well in this tradition of reducing religion, faith in God, and even spirituality to something merely and exclusively material, as if religion is only a neuronal or neurological phenomenon, a matter of feelings and emotions at best. One gets the impression that Hamer and his like-minded spirits just want this all to be true; they want religion to be nothing more than neuronal activities, so that religion can be discarded as being a mere hallucination, illusion, or delusion. However, when people like Freud claim that basic beliefs are the rationalization of our deepest wishes, wouldn't this also entail that their own atheistic beliefs could also be the rationalization of their own wishes? But materialism just cannot answer the big questions of life, in spite of the fact that this statement may cause a potential outrage to the gatekeepers of materialism. It always baffles me how materialism, for some people, has such a spiritual appeal to it.

How could there be truth to what science claims if everything were merely material? Just as there is no way for "true and false" or "right and wrong" to exist if *matter* is all there is, so there is no way for the "divine" to exist, other than as a neuronal phenomenon. Genes can never produce truths and untruths—certainly not the truths and untruths of religion. The best genes could ever produce for religion are neural activities that evoke feelings and emotions. For those who do not accept religion, that is all they need. But then we are no longer talking religion. Religion is not a brain issue but a mind issue.

The gene that Hamer discovered or perhaps invented—is certainly not a "god gene," it is not even a "religion gene," but at best a "spirituality gene" that may enhance spiritual feelings and mystic experiences. But religion is certainly not identical to those experiences; it may lead to them but does not depend on them; they are neither necessary nor sufficient for religion. Instead, religion is about a relationship—a relationship between Creator and creature (*religare* in Latin means something like "to reconnect"). It is belief in a factual truth—in the fact that we are "creatures under God."

By reducing religion to spiritual feelings and emotions, to neuronal or neurological activities, Hamer is saying that religion is ultimately not what it appears to be—which amounts to materialism in optima forma. But that is a philosophical stance, not a scientific discovery. If I had to give his gene a label, I think "hallucination gene" would be the best fit. In general I would say that Hamer did a disservice to genetics with his work.

Step three—believing in God is more than a matter of mere belief or spiritual feeling. Abstracting from what is behind and beyond religious beliefs, emotions, or experiences would be comparable to studying science as a (merely) sociological or psychological phenomenon—say, a set of convictions, activities, and problem solving rules—while ignoring, or even denying, that science is (also) about a reality beyond the system of science. In a similar way, religion also refers to something beyond a set of religious beliefs, feelings, experiences, and customs.

How could believing in God be more than a matter of mere belief or spiritual feeling? Well, in the previous three chapters, we discussed that free choices, rational decisions, and moral actions cannot come from our genes or DNA. So the question is where they do come from then. My answer is that their only real foundation can be found in the existence of a Creator God. Why do I think so?

If God is indeed omnipresent, "everywhere," God may seem to be "no-where." So God's omnipresence only makes for God's seeming absence. As they say about fish: The last thing a fish would ever discover is water. Something similar holds for God's existence. Yet, we have some very powerful pointers or indications of God's existence. Just think of the following series of (rhetorical) questions: How could nature be intelligible if it were not created by an intelligent Creator? How could there be order in this world if there were no orderly Creator? How could there be scientific laws if there were no rational Law-giver? How could there be design in nature, if there were no intelligent Designer? How could there be moral laws, if there were no moral Lawgiver? How could there be human minds, if the universe were mindless? How could there be human freedom if there were no God who has freely created us after his own image?

The answer to all the above questions is either "there is no explanation" or "there is only one explanation, God." Choosing the option that there is no explanation is basically an irrational response that leaves us stuck in a purely material universe. Take the law of gravity—it may help us explain things, but it cannot explain itself. Those who embrace materialism I would ask, how can matter ever explain itself, its own existence? It cannot just pop up out of nothing—nothing comes from nothing, as the saying goes. Since I do not see how matter could ever explain itself, a solely material universe is essentially an absurd universe. But I also realize there is just no rational way of changing the minds of people if they reject rationality.

Therefore, I think the other option is the only *rational* alternative: Only the existence of God can explain that there is a universe, that there is order in this universe, that this universe is intelligible, that there are laws of nature, and that there are moral laws. This is the only way we can take the world as something created according to an intelligible plan accessible to the human intellect through the natural light of reason. Because there is a Creator, we have not only a rational Lawgiver—who guarantees order, intelligibility, and predictability—but also a moral Lawgiver—who guarantees decency, integrity, conscience, responsibility, justice, and human dignity that comes with human rights. As Albert Einstein once put it, "Everyone who is seriously involved in the pursuit of science becomes convinced that a Spirit is manifest in the laws of the universe."

If we do not accept this answer, then we have basically given up on rationality and morality, for they would have become utterly baseless. Without God, scientists would fundamentally lose their *reason* for trusting their own scientific reasoning. Without the notion of the universe as a created entity, science would become a shaky and problematic enterprise. Without the notion of a Heaven with absolute and universal moral values and laws, everything would be permissible, and we would have no *right* to claim any moral rights. In short, if we lose the notion of creation, plus the trustworthy order it comes with, then natural laws and moral laws would become utterly questionable and problematic.

Let us take the laws of nature first—laws such as Newton's law of gravitation, his three laws of motion, the ideal gas laws, Mendel's laws, the laws of evolution, and so on. In a physical sense, laws are nowhere, and yet they are everywhere. Where do these laws come from? Not from the Big Bang, for they must have already been there when the Big Bang started, as it could not have begun without the law of gravity, for instance. Saying that they are just part of the "cosmic design" merely shifts the question. Even if physics could ever explain why the universe is the way it is, including its physical constants, we would still be left with the meta-physical question as to where the laws come from that explain all of this. Those who think that gravity would be able to "create" the universe all by itself use gravity as a magic wand that does not seem to need any further explanation.

Something similar holds for our moral laws. The following impartial testimonies can attest my point. Even an atheist such as the French philosopher Jean Paul Sartre realized that there can be no absolute and objective standards of right and wrong, if there is no eternal Heaven that would make moral values objective and universal. The German philosopher Friedrich Nietzsche was another atheist who clearly understood how devastating the decline of religion and a belief in God is to the morality of society, when he said that humanism and other "moral" ideologies shelter themselves in caves and venerate shadows of the God they once believed in; they are holding on to something they cannot provide themselves, mere shadows of the past. And the non-religious philosopher Jürgen Habermas expressed as his conviction that the ideas of freedom and social co-existence are based on the Jewish notion of justice and the Christian ethics of love.

No wonder then that the *United States Declaration of Independence* makes it very clear that we are endowed by our Creator with certain unalienable Rights—because they are *God*-given rights. President John F. Kennedy summarized this well in his Inaugural Address: "the rights of man come not from the generosity of the state, but from the hand of God." To put it differently, human rights are not *man*-made entitlements but *God*-given rights that we cannot invent and manipulate at will. We need a transcendent authority to sanction and recognize right from wrong.

Some people might object that morality cannot be based on God. In contrast, I would argue that those who say God cannot be the basis of morality, because there are so many different religions, commit the same fallacy as those who say evolution cannot be the explanation of the diversity of life because there are several theories as to how life unfolded. Instead I would maintain that, without God, anything is permissible—or at least we can make it permissible by autonomously changing absolute moral values into our own relative moral evaluations. Without God, even an oath becomes meaningless.

The conclusion of all of this is that there must be someone behind and beyond all we see and experience—and this "someone" is the Creator of Heaven and Earth. If you would ask me, are we alone in this universe, my answer would be, we are not. I am not stating here that there are also other forms of life in this universe. Perhaps there are. But even if there are, we would still be alone... unless there is a God.

Religion is about a personal relationship between Creator and creature. It is more than mere belief—it is a logical necessity for our world to exist and for our rational and moral claims to become valid. Religion is not just about feelings and experiences we have on our side of the relationship, but it is also about a reality called God that we encounter on the other side of the relationship. Just like Mendel and the first geneticists explained what they saw with what they could not see, so we too should try to explain what is seen, the creation, with what is unseen, the Creator.

In other words, religious belief is not merely a belief that is credible, believable, or acceptable to a certain degree, but it is the very basis on which our rationality and morality, all our thinking and doing, are grounded—whether we are aware of this or not. Without this religious belief, rationality and morality would be baseless. Apparently, we need some beliefs so that we may understand. Albert Einstein was right when he said that science without religion is lame (but he also rightly added that religion without science is blind). Believe it or not, it is faith in a Creator that makes us understand, turning science into a faith-based enterprise—based on a belief in a Creator God who has given this world its order and its laws, so we can explore, understand, explain, and predict. He created a "trustworthy" universe.

Therefore, denying that there is a Creator is actually an acid eating away not only the foundation of science, but of all rationality and all morality. Our universe must somehow possess an "inner logic" accessible to human reasoning. At the very moment we claim that this belief is merely a product of genes, we have undermined all our own activities, scientific and otherwise. Hence, *genetics cannot explain religion or any religious beliefs, for it must assume certain religious beliefs before it can even start its explanatory job.*

I am not saying this is all there is to religion, but I am saying that this is at least an important part of religion. Hamer and his like-minded colleagues wish to deny this and reduce religion to a series of self-transcending feelings and mystical experiences, thus missing the core of what they are trying to study and explain. To say that some gene creates "religious or spiritual experiences" entirely misses the point as to what religion is all about. Religion is not about those experiences per se, but about what they reveal to us from the "other side." Religion is not about hallucinations, delusions, and illusions, but about a real, yet unseen, heavenly world—about something or someone beyond this world. It is not merely about feelings and experiences but also about certain beliefs as to what is true or false and what is right or wrong in this universe.

Let me give you one specific example to end with. Religion is one way of coping with our knowledge that death is inevitable. Religion diminishes the hurt of death's certainty and permanence and the pain of losing a loved one with the promise of reuniting in another life—not as a form of wishful thinking but as a pillar of religious faith, not as a product of genes but as a given captured by the human mind. Humanity has always known this—intuitively if you will—from its very beginning. In early primeval history, we find already evidence of elaborate burials; such burials with grave goods indicate a belief in an after-life, for the goods are there because they are considered useful to the deceased in their future lives.

How different this is from female apes that continue carrying their dead newborns around for quite a while, without having any idea of what is happening until at last they give up and drop the dead remains of the baby. In Guinea, West Africa, for instance, chimpanzee mothers were seen in nature carrying and grooming their offspring's lifeless bodies for up to sixty-eight days. By the time the corpses were finally abandoned, the bodies had mummified and developed an intense smell of decay.

Believe it or not, evolutionists gave this observation a peculiar twist: These they said must be sixty-eight days of actually mourning the dead! All I can say in reply is that only human beings could come up with such an explanation. I would consider this sixty-eight-day period not a time of "mourning," but rather a time of "ignorance." Besides, I would ask such evolutionists where the burial and the grave are once this period of "mourning" was over.

Apparently, only the "finite" human mind is able to catch a glimpse of the Infinite! This capacity of the human mind is rightly called self-transcendence—referring to the transcendence of something or someone more than our own selves, and certainly not in the sense Hamer uses that term. Man's drive for self-transcendence can be found even in the earliest archaeological data, such as art and burial rituals. Blaise Pascal—the mathematician, physicist, and philosopher—used to speak of a hole in each one of us, an "infinite abyss that can be filled only with an infinite and immutable object; in other words by God himself." I do not think this hole is programmed in our genes—and neither is religion.

17. When Did It Start?

We discussed in the previous three chapters that human beings are part of all living beings, but unlike other living beings, they are rational, moral, and religious beings—at least potentially so. Humans are animals, but animals with a difference. Rationality, morality, and religion seem to set us apart as human beings, because these characteristics do not come from the animal world and are not etched in our DNA. Only the human *mind* has access to the realms of rationality, morality, and religion.

So we end up being very different from our "relatives" in the animal world. Although we are flesh as they are flesh, because we breed, feed, bleed, and excrete like they do, we are also very exceptional creatures in this universe, endowed with the capacity for rationality, morality, and religion. We have got to be more than what science tells us, otherwise science would in fact not have much to tell us. In addition to our five material senses, we have three immaterial senses—the sense of "true or false," the sense of "right or wrong," and the sense of "material or spiritual."

This leads us almost automatically to the question as to when humanity came along and received these immaterial "senses." Scientists have always been intrigued by this issue. We discussed already that geneticists have traced our human origin back to "mitochondrial Eve" and "Y-chromosomal Adam" (see Chapter 11), but we also found out that those concepts are more of a mathematical than a physical nature. They do not represent the first beginning of humanity, but they merely stand for a construct of our two most recent common representatives in a matrilineal or a patrilineal way.

Paleontologists had tried something similar already much earlier. They came up with certain anatomical characteristics that are supposed to uniquely define human beings. One of the earliest criteria was brain volume. When the brain volume of a skull was found to be above a certain minimum—some said 1,000 cc—they were ready to assign the label *Homo sapiens*. This is a rather arbitrary criterion, though. *Homo sapiens* is indeed the "rational man" who surely has a relatively large brain compared to other mammals, yet brain volume can vary widely in human beings. Besides, brain mass or quantity is not clearly correlated to brain quality, let alone mental activities. Neanderthals score relatively high on a brain volume scale—but so do elephants. And finally, I would add we are confusing the mind with the brain again.

Next they declared that each time a fossil showed signs of being capable to walk upright on two legs (biped), we should consider some kind of human-like qualification. When fossils showed evidence of bipedalism, they were assigned to the species *Homo erectus* (upright). Sure, being a two-footer seems to be a big plus for other human features— the main advantage being that the neck muscles put less restraint on the expansion of the brain part. Although humans do stand on two legs, their skeleton was originally designed for four.

Evolution accomplished this transition by breaking up the bow-curved arch of the back of four-legged animals into the s-curve we as bipeds now possess—leading also to a more basin-shaped pelvis, good for balance but bad for birthing because it narrows the birth canal. But the question remains whether being a biped is really a decisive factor for being human. It may not even be a necessary condition for being human, let alone a sufficient condition. Many animals are bipeds; all birds, all kangaroos, most dinosaurs, and even chimpanzees have been able to walk or hop pretty well on two feet. Does that make them come closer to us?

The next step was that any kind of evidence of tool-using and tool-making was taken as another important indicator of human qualifications. So they assigned certain fossils to a new species, the species *Homo habilis* (the "handy" man). Supposedly, the ability of using tools is facilitated by a so-called opposable thumb, which is a pad-to-pad kind of precision grip. It is not unique to humans, though, and can be found to different degrees among other Primates such as baboons; even opossums, koalas, and pandas have some form of "opposable thumbs."

Making tools, on the other hand, requires additional features such as a certain brain capacity. However, there is more and more evidence showing that chimpanzees and some other animals not only are able to use tools but also have some capability of making tools. Did humans therefore lose another unique human characteristic? I do not think so. Perhaps tool making is a necessary but not sufficient condition for being a "rational designer." The human mind of a tool maker is able to create tools according to some mental *concept* of a tool object with which they constantly compare the real product they are working on. Thus they anticipate the future by means of abstract rational conceptions residing in their minds. We discussed earlier that concepts are very different from images (see page 135).

There might be even one more step. Some claim that what makes humans unique is their use of language—which would entitle us to be called *Homo loquens*. Unfortunately, skeletons do not show that feature. Besides, language may be an important tool for the human mind, but it is just a tool like the brain is a tool for the mind. Even if language use has a genetic basis, that does not make the mind a genetic issue. Yet, there are some vague indications that there are language genes. In 1990, geneticists at the Institute of Child Health in London first reported a speech disorder that appeared in three generations of Britons known as the KE family. They found 15 affected family members who seemed to have inherited problems with grammar, syntax, and vocabulary that were tied to poor control of facial muscles and difficulty pronouncing words. Although it seemed clear that there had to be a genetic link, researchers hunted for more than a decade before they could identify the gene responsible.

It was finally found in a fetus that had a translocation in chromosome 7, and later developed speech and language problems strikingly similar to those seen in the KE family. The translocation found in the boy had disrupted a gene called *FOXP2*, and they found the same mutation in the 15 members of the KE family. It is clear that *FOXP2* does not single-handedly wire the brain for language. It is rather a transcription factor, turning other genes on or off by telling them whether to transcribe their DNA into messenger RNA, which then leads to the production of certain proteins. As a result, *FOXP2* has a broad repertoire in embryonic development, playing critical roles in the formation of the lungs, heart, and intestines. So do we have a "language gene" here? Certainly not—it does not create the ability for language, yet it is somehow involved in the molecular pathways behind speech and language. (But so do genes for hemoglobin; if those do not work properly, speech will stop too.) Obviously, the discovery of the *FOXP2* gene yields more questions than answers.

What then is the problem with all the previous attempts to locate the exact spot where humanity came into existence? Scientists can only work with physical or material characteristics. In that specific sense, Darwin was right when he said that the difference between animals and humans is one of degree, not kind. We are made of their cloth, so during evolution there must have been some features in the animal world that came closer and closer to features more characteristic for the human world—features such as making tools, walking upright, having a relatively large brain volume, and using language. In a purely biological sense, there is only a very gradual transition; we surely are "glorified animals" and animals are "humans-in-the-making."

And yet, no species is more unlike its ancestors than the human species. Not only does analysis of the human genome show that chimpanzees are two to four times more genetically diverse than the global human population—which makes humans unique as a species in that we are all extremely closely related to each other, much more so than even our closest living relatives—but there is also another dimension to human uniqueness. Genetic, physiological, and anatomical features are not the determining factors for being human. Although we do breed, feed, bleed, and excrete more or less like they do—true, we are connected all the way down—what does set us apart from the animal world is our capacity for rationality, morality, and religion, so we found out. Well, those features do not come from our genes, our anatomy, or our physiology. So how could we ever "diagnose" them by merely looking at skulls and skeletons, or even DNA? In other words, the question of whether Neanderthals were human beings has nothing to do with their "looks," but depends on whether they were rational, moral, and religious beings—and obviously we have no way of knowing, other than going by burial practices or written records (but the latter came much later).

No wonder, archeologists have considered evidence of burial rituals and the presence of grave goods an important indicator of being human, because these may signify a concern for the dead that transcends daily life. If so, we could speak of *Homo religiosus*. However, even burials are hard to diagnose. There are serious questions, for instance, as to whether Neanderthals practiced burial rituals. The earliest undisputed human burial dates back 90,000 years, in the Skhul cave at Qafzeh, Israel.

So a search for the "real" Adam and Eve, and for the emergence of the human mind, is on—but it is beyond the scope of science, because the human mind, unlike the human brain, is the subject of science but not its object. It always baffles me how evolutionists like to downgrade the human mind while touting their own minds. I keep stressing that it requires a mind to study the brain. But those who have been trained enough to watch only for gradualness and continuity in evolution may not notice anymore how much the human race differs from the animal kingdom, because the physical similarities are so major and obvious indeed that they make the non-physical differences appear minor or even vanish from sight. Nevertheless, we can only identify human beings by their minds, but the mind eludes us when it comes to skeletons and DNA. Although there is something like neuro-genetics, there is no such a thing as mind-genetics.

I can already hear some protest vehemently: This is not science! They are right—this is not science. But does it mean that what I said is not true? If you think it is not true because it is not scientific, then you are probably a supporter of what is called *scientism*, which is a world-view, or actually a "dogmatic creed," stating that science provides the only valid way of finding truth, thus eliminating everything that cannot be counted or measured.

Supporters of scientism claim that "the real world" is only a world of quantified material entities. They pretend that *all* our questions have a scientific answer phrased in terms of particles, quantities, and equations—or in this specific case, of genes and fossils. Their claim is that there is no other point of view than the "scientific" world-view. Their trust is entirely in science, without trusting anything else, since they only acknowledge one territory—the territory of science. They believe there is no corner of the universe, no dimension of reality, no feature of human existence beyond its reach. They have a dogmatic, unshakable belief in the omni-competence of science.

Curiously enough, those who defend scientism maintain that science is the only way of achieving valid knowledge, but they seem to be unaware of the fact that scientism itself does not follow its own rule. How could science ever prove all by itself that science is the only way of finding truth? There is no experiment that could do the trick! Science cannot pull itself up by its own bootstraps—any more than an electric generator could run on its own power. So the truth of the statement "no statements are true unless they can be proven scientifically" cannot itself be proven scientifically. Scientism is in essence an immaterial and self-refuting claim about the material world—once we consider it to be true, it becomes false. It is not a scientific discovery but at best a philosophical or metaphysical stance.

To best characterize such an attitude I like to borrow an image from the late psychologist Abraham Maslow: If you only have a hammer, every problem begins to look like a nail. So I would suggest to not idolizing your "scientific hammer," because not everything is a "nail." Even if we were to agree that the scientific method gives us better testable results than other sources of knowledge, this would not entitle us to claim that only the scientific method gives us genuine knowledge of reality. Since science has no scientific test for world-views, we have another "boomerang theory" here: *If I believe that science is the only valid source of knowledge, then my belief cannot be validated by science, so I do not have to believe that science is the only valid source of knowledge.*

Then there is another problem with scientism. It is an ideology that has us shackled in a physical, material world. However, the claim that "physical matter is all there is" necessarily implies that this very claim does not and cannot exist because claims are essentially *not* physical; so if we think that such a non-physical claim does exist, there must be more than physical matter in this universe. *Those who deny the existence of things immaterial also deny the existence of their own immaterial denial—as well as all their scientific claims.* How could there be truth to what science claims if everything were merely material? Scientism is in fact self-defeating; it declares everything outside science as a despicable form of metaphysics, in defiance of the fact that all those who reject metaphysics are in fact committing their own version of metaphysics. Metaphysics may be a "dirty word" to some, but we all are surrounded by it—like it or not.

In short, there is so much more to life than science, for there is so much more that "counts" in life than can be counted. We should never forget that there are so many aspects of life that are off-limits for science—and the mind issue may be one of them. Science just cannot account for all that needs to be accounted for. The late UCB philosopher of science Paul Feyerabend came to the right conclusion when he said that science should be taught as one view among many, and not as the one and only road to truth and reality. The astonishing successes of science have not been gained by answering every kind of question but precisely by refusing to do so.

As a result, scientific knowledge is not a superior form of knowledge; it may be more easily testable than other kinds, but it is also very restricted and therefore requires additional forms of knowledge. Consider this analogy: A metal detector is a perfect tool to locate metals, but this does not mean that the world is only made of metals. That is exactly where scientism goes wrong: Instead of letting reality determine which techniques are appropriate for which parts of reality, scientism lets its favorite technique dictate what is "real" in life—in denial of the fact that science has purchased success at the cost of limiting its ambition.

So I think we have many reasons to abandon and dump scientism. Going back to our original question, the fact that science has found a genetic "Adam" and "Eve"—representing our most recent genetic ancestors—does not mean we have to stop searching for the "real" Adam and Eve, who stand at the origin of rationality, morality, and religion. But such a search is beyond the scope of science.

V. Conclusion

Through this entire book, I have seriously tried to show the power of genetics as well as its limitation, and I hope I did so in a fair and balanced way.

Where does its power come from? The power of genetics is that it has managed to reduce a "complex entity" to its "simpler parts." It reduces a population to a series of organisms; it reduces an organism to a series of genes; it reduces genes to a series of DNA nucleotides. I call such an approach *methodological* reductionism. It is a very powerful methodology that simplifies reality by reducing the complexity of the original to a manageable model related to a soluble problem.

The history of genetics testifies to the power of this technique. It is a "piecemeal" approach that scientists excel in by dissecting things into pieces. "Good" scientists are those able to demarcate their area of investigation, able to limit themselves to factors that are relevant to their object of investigation, and capable of eliminating factors that might interfere with their search by keeping them under strict control. "Dissect!" is the motto that drives good science. Mendel was good at it—and so were his successors. They dissected organisms into cells, cells into chromosomes, chromosomes into genes, and genes into nucleotides.

Is there any limitation then? The limitation of genetics is that it cannot answer *all* our questions, not even all our genetic questions. People who do not acknowledge this limitation commit themselves to another form of reductionism—*ideological* reductionism. According to this ideology, genetics is a know-it-all; because it has been so successful with its methodological reductionism, it must presumably be successful in everything.

I doubt whether this latter claim is true. Whereas methodological reductionism reduces the complexity of the whole to the simplicity of its physical parts, ideological reductionism goes one step farther—or actually too far—by claiming there is nothing else but those simpler parts. It claims that the complexity of an organism is the mere result of the simplicity of its components. *Neglecting* what is outside one's scope may be a wise scientific strategy, but *denying* what is outside one's scope turns a good strategy into an unwarranted, poor ideology. Once we have been squeezed into the models of genetics, we supposedly end up being "nothing but" a collection of genes or a string of nucleotides. That is no longer science, but rather a (poor) kind of world-view—actually megalomania. If we are literally nothing but DNA or a bunch of genes, then we are worth nothing more than a fragile molecular structure, and so is all we pretend to know.

In this book, I have taken methodological reductionism very seriously—for it is a powerful tool—but I have also protested vehemently whenever methodological reductionism is being turned into ideological reductionism. It might be okay, for scientific purposes, to reduce a human being to genes and DNA molecules, but that does not mean human beings are ultimately "nothing but" genes and molecules. Interestingly enough, the power of genetics—reducing the complexity of reality to the simplicity of genetic models—is at the very same time also its weakness of no longer being able to do justice to the complexity behind its simplified models.

You might still be questioning, though, whether there are really any limitations to science. Not surprisingly, my answer is in the affirmative: Science has many limitations. Because all sciences use their own models, they are "blind" for what does not fit into their models. Each model is based on its own assumptions and refers to its own kinds of causes and boundary conditions. Each model is only a surrogate for "the real thing." The only model that could ever qualify as a perfect replica of the original is the original itself. Therefore, scientists of the different areas or fields of science have a very selective approach; everything outside their scope is on their "blind spot," because they *neglect* what they did not *select*. Physicists, for instance, only use a "physical eye" to capture the physical parts of this world; chemists have a "chemical eye"; and geneticists see everything with a "genetic eye." But let us not forget that physics cannot capture everything, neither can genetics. Arguably, even all sciences combined cannot capture all there is, for they only capture what can be measured and counted.

So what I just said about the different areas of science also applies to science in its entirety. Science in general is "blind" for all non-scientific, immaterial, and spiritual aspects of life; it is "blind" for everything that cannot be counted, measured, or quantified. Failure to distinguish the immaterial from the material made Duncan MacDougall, an early 20th-century physician in Haverhill, MA, measure how much mass a human body would lose when the soul departed the body upon death. He came up with a weight of 21 grams!

To gain access to the huge domain of all that counts but cannot be counted, measured, or quantified, we need more than a "scientific eye." Let me mention a few examples. Just like the "physical eye" sees colors in nature, so the "artistic eye" sees beauty in nature, the "rational eye" sees truths and untruths, the "moral eye" sees rights and wrongs, and the "religious eye" sees a spiritual dimension in life. All these different "eyes" are in search of reality, but each one "sees" a different aspect of it. Even astronomers do not deal with the universe in its entirety, but only with its physical aspects.

As a consequence, there are also many kinds of blindness: Not only can one be physically blind, but also morally blind (when unable to see what is right and what is wrong), or rationally blind (when unable to see what is true and what is false), or even spiritually blind (those who fail to see the Heaven beyond the Earth). We all have our blind spots and weak spots, but we could at least try to compensate for them.

Let me put this differently. Reality is like a jewel with many facets; you can look at it from various angles, with different eyes, from different perspectives. As a consequence, physicists perceive primarily physical phenomena, whereas biologists see mainly biological facts—but all of them are by definition "blind" for other aspects of reality. Even what they call the "scientific world" is just one particular aspect of the "real world." The fact that other aspects are beyond the scope of science does not make them less "real" or less "factual" or less "objective" or less "valid." What you choose to neglect you cannot just reject.

Because of all of the above, we can say again that genetics answers many of our questions, but that it does not answer *all* our questions. It may not be able, for instance, to answer the questions we have about free will, rationality, morality, and religion. In this book, I tried to show that it is in fact even pointless to go in pursuit of genes for our free will (or lack thereof), of genes for our human capacity of rationality, of genes for our human capacity of morality, and of genes for our human capacity of religion—in spite of claims often made to the contrary. Much in our *bodies* may be controlled by genes, but that does not mean our *minds* are under their control too. Genes do not control the free choices we make, the truths we uphold, the rights we claim, or the religious beliefs we hold. I hope I made that convincingly clear.

To those who still keep claiming that *everything* is determined by genes—including rationality, morality, and religion—I can only say that there must also be then an allele for making such a claim. Obviously I do not have such an allele, if there is one.

VI. Index

A

a priori.... 111, 112, 114, 115, 136, 141
ABO system 9, 19
abortions.. 60
achondroplasia............................. 19
activator 74
adaptation..................................... 51
addiction32, 121, 122, 123, 124, 125, 163
   behavioral 124
   substance 122
adoption research 30
alcohol....................... 21, 32, 121, 123
alcohol dehydrogenase (ADH)21, 105, 123
alcoholism 24, 123
aldehyde dehydrogenase (ALDH). 123
allele 20
allele and gene 4
altruism 154, 163
Alzheimer's disease 17, 104
amino acid..................................... 72
amygdala..................................... 124
amyloids..................................... 104
anthropomorphism................134, 151
antibiotics..................................... 91
antibody 10
antigen............................... 9, 10, 19
antioxidants 88
apolipo-protein AI........................ 90
Aristotle 129
ascorbic acid................................. 75
autism................................ 17, 22, 27

B

Bateson, William 1, 3, 4, 5, 71
Becker muscular dystrophy (BMD) 14
behavior 22, 28, 33, 34, 119, 121
   altruistic.................................... 155

   social...................................... 144
behavioral genetics 22, 163
Big Bang....................................... 170
bipolar disorder 27
blood clotting................................97
blood clotting factor 73
blood types.................................... 9
Bombay phenotype........................10
boomerang theory.......... 139, 142, 181
brain 117
brain vs. mind 116, 128
Brenner, Sidney............................113

C

cancer 86, 88
carcinogen.....................................86
carrier 6, 12, 13, 14, 34, 42
catalyst.................................... 13, 16
cause and effect............................120
central dogma.................................72
chromosomal Adam....... 101, 102, 175
chromosome
   and genes................................. 5
   map....................................... 6
   Philadelphia..............................84
   sex 8, 14
   silencing...................................62
chromosomes 60, 62, 63
CNVs....................16, 17, 30, 95
codon 72, 85, 89
coenzyme 105
coenzyme A 108
Collins, Franics 79, 164
color blindness 14, 25, 98
color vision 98
conception............................. 63, 65
contradiction 115
copy number variants..........See CNVs
Correns, Carl............................. 1, 4
creation 170, 172

Creutzfeldt-Jakob disease............ 104
Crick, Francis ... viii, 71, 106, 140, 141
crossing-over .... *See* recombination
   unequal..................................... 94

D

Darwin, Charlesv, 51, 52, 53, 54, 55,
   56, 57, 58, 59, 138, 140, 141, 178
Dawkins, Richard 167
DDT... 91
De Duve, Christian 108
De Vries, Hugo............................... 1
denaturation.................................. 79
depression....................23, 32, 125
design
   biological vs. cosmic 56
   cosmic 107
determinism....................23, 32, 114
   genetic.............................. 22, 23
diabetes 34
DNA 71, 72, 73, 74, 85
   direction.................................. 72
   double helix 72
   duplication 94
   intergenic................................. 93
   junk 93
   marker 101
   mitochondrial (mtDNA) 99, 110
   non-coding........................ 75, 93
   polymerase......................80, 105
   regulatory................................. 74
   repetitive 75
   replication..........................36, 103
   selfish 106
   sequencing............................... 79
   signature 100
   transcription............................. 72
   Y-chromosomal........................ 99
DNA methylation........................ 105
dominant 2
dopamine........ 34, 122, 123, 124, 161
Down syndrome........... 60, 61, 68, 84
Duchenne muscular dystrophy (DMD)
   14
dwarfism 19
dystrophin............................ 15, 73

E

Edwards syndrome 61
egg cells.......................................60
Einstein, Albert131, 136, 137, 170,
   172
electroencephalogram (EEG) 127
electropherogram...........................80
enhancer............................. 74, 76, 90
enzyme 13, 16, 72, 109
epigenetics 33, 35, 36, 105
ethics 144
eugenics.......................................69
evolution...............................44, 75
exome ....iii, 81

F

facts...138
falsification 136
fertility...62
Feyerabend, Paul 182
fitness 41, 42, 50, 51, 52, 96
   D- 53
   F- 52
formaldehyde 86
FOXP2...178
frame-shift 86, 94
free radicals............................ 82, 88
free will 113
Freud, Sigmund...........................167
functional   magnetic   resonance
   imaging (fMRI) 128
functionality.......... 50, 59, 66, 74, 136

G

gay gene...................24, 26, 163, 164
gender vs. sex...............................28
gene.............................. 20, 22, 73
   and chromosome........................ 5
   definition........................... 7, 8, 76
   dosage 17
   duplicate..................................96
   duplication................... 16, 91, 94
   expression 17
   fusion 99
   gay-....................24, 26, 163, 164
   god-....................160, 163, 164

history 5, 71
hox or homeobox 75
hybrid 99
jumping 95
linkage 5, 6
mismatch-repair 83
mono-factorial 20
multi-factorial 20
new 75
oncogene 85, 86
overlapping 73, 77
proto- 97
pseudo- 74, 95, 96
tumor-suppressor 86
vs. allele 8
gene and allele 4
gene therapy 64
gene-pool model ..... vii, 40, 41, 43, 44
genetic determinism . 22, 113, 114, 158
genetic drift 41
genetic marker 25
genetic vs. hereditary 33, 50
genome
    map 79
    vs. exome 81
    vs. genotype 35
genomics 81
genotype 9
    vs. genome 35
    vs. phenotype 35, 49
Gilbert, Walter 113
god gene 160, 163, 164

H

Habermas, Jürgen 171
Haldane, J. B. S. 118, 155
Hamer, Dean 24, 25, 26, 27, 160, 161,
    162, 163, 164, 165, 166, 167, 168,
    173
haplo-group 101
haplotype 100
    map 101
Hardy-Weinberg law 40
hemoglobin 19, 41, 95, 105
hemophilia 14, 25, 97
hereditary vs. genetic 33
heterozygote 3, 42
heterozygous 4

histone 105
homeobox gene 75
Homo erectus 176
Homo habilis 176
Homo sapiens 176
homosexuality .. 22, 24, 25, 26, 28, 163
homozygote 3, 42
homozygous 4
Hox gene 75
Human Genome Project . 7, 78, 81, 164
Huntington's disease 11
hybridization 2, 3

I

immaterial 115, 128
incest 153
induction
    generalizing 111
infinite regress 114
information 117
insulin 104
intellect vs. intelligence 131, 134
intelligence vs. intellect 131, 134
intelligibility
    as a presupposition 137
intron 73, 77, 81

J

Johanssen, Wilhelm 5

K

Kell system 9
Klinefelter syndrome 61, 84
Kölreuter, Joseph 2
Kuhn, Thomas vi

L

lactose intolerance 90
Landsteiner, Karl 9
language 177
laws of nature 57, 107, 120, 130, 135,
    146, 149, 169, 170
Leibniz, Gottfried 127
Lewis system 9
limbic system 122, 124

Lincoln, Abraham......................... 149
linkage.. 5, 6
linkage map................................... 79
Linnaeus, Carl 2
locus control region (LCR)............. 96

M

magnetic resonance image (MRI) . 127
malaria 42, 91
materialism.................. 166, 168, 169
Medawar, Peter.......................39, 112
melanin 20
Mendel, Gregor1, 2, 3, 4, 5, 6, 39, 40,
    49, 71, 170, 172, 183
menstrual cycle.............................. 62
micro-array technology 95
mimicry.. 53
mind... 22
mind vs. brain.......................115, 128
miRNA... 77
mitochondrial Eve......... 101, 102, 175
mitochondrium 99, 110
model .... vi, 23, 39, 43, 44, 184
    gene-pool................................. 40
    in mimicry 53
    optimality................................ 51
molecular clock.....................100, 101
mono-amine oxidase (MAO).......... 34
morality................................144, 170
    not in the animal world............ 150
Morgan, Thomas Hunt6, 7, 8, 25, 71,
    76, 78, 79
mosaic syndromes.......................... 63
mRNA................................See RNA
MSN system 9
murder... 153
muscular dystrophy............ 14, 15, 73
mutagen.. 86
mutation 82
    at the chromosome level 84
    at the gene level 85
    beneficial 89, 96
    eandom 82
    frame-shift 86, 94

N

natural selectionvii, 40, 49, 50, 51, 52,
    53, 55, 56, 58, 59, 97, 107, 109,
    122, 138, 139, 141, 142, 153, 155,
    162
    of molecules............................108
    survival of the fittest 55
nature-nurture debate........ 30, 35, 125
neuro-science 127, 140
Nietzsche, Friedrich....................171
nucleotides..................................71

O

obsessive-compulsive disorder......121
oncogene................................86, 94
    viral.......................................85
operator.......................................74
operon...78
optimality model 51, 52
order
    as a presupposition........... 135, 169
osteoporosis 90
ovaries 62

P

Paley, William.........................55, 57
paradigm............................ vi, 2, 127
paradox................ 114, 118, 138, 142
Patau syndrome 61
paternity test................................10
Penfield, Wilder 119
penicillinase 91
phenotype 9
    vs. genotype 4
phenylketonuria............................20
Philadelphia chromosome 84
phospholipids 13
plasmid..................................64, 79
polymorphism 42, 123
    balanced..................................42
population genetics........................40
predisposition...........................21, 32
pregnancy 60, 62
primer...79
prion..104
progeria.......................................85

prokaryotes 110
promoter 74, 78, 85, 90
protein 72, 73
puberty 62

R

radiation 87
  ionizing 87
  non-ionizing 87
radicals .... See free radicals
radioactive isotope 87
radio-nuclide 87
random 68, 82
  vs. fate 68, 83
Randomness 67, 82, 83
rationality 129, 135, 170
  not in the animal world 131
recessive 2
recombination 6, 8, 9, 25, 76, 78, 94, 99
reductionism
  ideological 183, 184
  methodological 183, 184
religion 22, 160, 164, 165, 168, 173
repressor 74
retro-transposon 36, 95
retrovirus 36, 85, 93
reverse transcriptase 36, 93, 95, 110
Rhesus factor 10
Rhesus incompatibility 10
Rhesus system 9, 10
ribozyme 109
RNA
  editing 77
  micro- 74, 77, 94
  miRNA 74, 94
  mRNA 72
  polymerase 78
  replicase 110
  rRNA 77, 94
  tRNA 72, 77, 94, 109
  virus 36, 93, 109

S

Sartre, Jean Paul 171
schizophrenia... 17, 22, 23, 27, 33, 163
scientism 180, 181, 182

self-control 125, 126
self-determination 121
self-reference 114
sequencing
  dideoxy 79
  genome vs. exome 81
sex vs. gender 28
Shaw, G.B. 57, 66, 69
sickle cell anemia 19, 41
silencer 74, 76
single-nucleotide polymorphism ... See SNP
SNP 85, 90, 100
socio-biology 155
species 2, 3, 44, 46, 47, 53, 74, 79, 153
sperm cells 60
splicing 73, 81
  alternative 73
  trans- 77
statistical research 31
styrene 86
survival of the fittest 67
Sutton, Walter 5

T

Tay-Sachs disease 11, 12
teleology 56, 57, 58, 136
  vs. causality 58
telomerase 36
thio-ester 108, 109
Thomas Aquinas 129
transposon 95
  retro- 36
Triple-X syndrome 62
trisomy 60, 61, 63
tRNA .... See RNA
tumor-suppressor gene 86
Turner syndrome 62, 84
twin studies 163
twins
  adopted 30
  identical 20, 25
tyrosinase 20

U

upstream 73, 78, 90
UV-rays 88

V

vector 64
violence.. 23
vitamin C...................................... 105
Von Tschermak, Erich...................... 1

W

watchmaker 55, 57
Watson, James.........71, 106, 140, 141
wild type 41

Wilson, E. O................................. 155
Wittgenstein, Ludwig 117

X

X-chromosome......................... 61, 62
XIST... 62

Y

Y-chromosome.............................. 61

www.ingramcontent.com/pod-product-compliance
Lightning Source LLC
Chambersburg PA
CBHW051646170526
45167CB00001B/355